Introduction to Algorithms and Data Structures in Swift 4

KÁROLY NYISZTOR

Copyright © 2018 Károly Nyisztor

All rights reserved.

ISBN: 9781973291749

DEDICATION

To my wife Monika and my children Ervin, Lea and Lisa.

CONTENTS

1 Introduction — 1
   1.1 Prerequisites — 2
   1.2 Why Should You Learn Algorithms? — 3
   1.3 What's Covered in this Book? — 4

2 The Big-O Notation — 6
   2.1 Constant Time Complexity — 8
   2.2 Linear Time Complexity — 13
   2.3 Quadratic Time Complexity — 17
   2.4 Hints for Polynomial Time Complexity — 21
   2.5 Logarithmic Time — 22
   2.6 Summary — 24

3 Recursion — 25
   3.1 What's Recursion? — 26
   3.2 How Does Recursion Work? — 29
   3.3 Recursion Pitfalls — 31
   3.4 How to Avoid Infinite Recursion? — 34

4 The Power of Algorithms — 37
   4.1 Calculate Sum(n) — 39
   4.2 Pair Matching Challenge — 42
   4.3 Find the Equilibrium Index — 46
   4.4 Summary — 50

5 Generics — 51
   5.1 Why Generics? — 52
   5.2 Generic Types — 54

5.3 Generic Functions	56
6 The Built-In Swift Collection Types	60
6.1 The Array	61
6.2 Accessing the Array	64
6.3 Modifying the Array	66
6.4 The Set	72
6.5 Accessing and Modifying the Set	75
6.6 Set Operations	78
6.7 The Hashable Protocol	81
6.8 The Dictionary	84
6.9 Creating Dictionaries	85
6.10 Heterogeneous Dictionaries	86
6.11 Accessing & Modifying the Contents of a Dictionary	88
7 Basic Sorting	90
7.1 Selection Sort	92
7.2 Insertion Sort	98
7.3 Bubble Sort	107
8 Advanced Sorting	113
8.1 The Merge Sort	114
8.2 Quicksort	125
9 Where Do You Go From Here?	130
9.1 Resources to sharpen your skills	131
9.2 Goodbye!	132
9.3 Copyright	133
About the Author	134

CHAPTER 1
INTRODUCTION

Thank you for buying my book *"Introduction to Algorithms and Data Structures in Swift 4"*. This book is going to teach you fundamental knowledge about algorithms and data structures.

SECTION 1.1
PREREQUISITES

This book is beginner-friendly. Prior programming experience may be helpful, but you need not have actually worked with Swift itself.

To implement the exercises in this book, you'll need a Mac with macOS 10.12.6 (Sierra) or newer. Sierra is required because Xcode 9 won't install on prior versions of macOS.

You'll also need Xcode 9 or newer. You can download Xcode for free from the Mac App Store.

We're going to use modern *Swift 4* to implement the source code in this course.

Swift 3.0 has brought fundamental changes and language refinements. Swift 4 added some useful enhancements and new features. All the samples are compatible with the latest Swift version. I am going to update the source code as changes in the language make it necessary.

The projects are available on Github, and you can download them from the following repository: https://github.com/nyisztor/swift-algorithms

The Swift language is available as open source. Visit http://swift.org for everything Swift related.

Also, you can visit my website http://www.leakka.com and my Youtube channel https://www.youtube.com/c/swiftprogrammingtutorials where you can find many Swift-related articles and video tutorials.

SECTION 1.2
WHY SHOULD YOU LEARN ALGORITHMS?

Computer algorithms have been developed and refined over the last couple of decades. The study of algorithms is fundamental to any programmer who plans to develop a software system that is scalable and performant.

Once we got past the basic *"Hello World"* beginner applications, we begin to realize that complex applications require a different approach. The software which used to work nicely during our tests becomes incredibly slow and frequently crashes in real-world situations. The reason is that we haven't prepared our system for real-life usage: while it ran without issues with small datasets during our tests, it fails miserably when the reality kicks in.

Algorithms are indispensable to building software that is capable of managing large amounts of data or solving complex problems efficiently.

Besides, you'll probably come across algorithm and data structure related questions during job interviews.

SECTION 1.3
WHAT'S COVERED IN THIS BOOK?

First, we'll talk about the Big-O notation, which is a mathematical model for measuring and classifying algorithms. It estimates the time efficiency of running an algorithm as a function of the input size. We are going to talk about constant time, linear time, polynomial time, and logarithmic time. To understand these concepts, we're going to implement, analyze and compare a various algorithms.

The next chapter is about recursion. Many algorithms and data structures rely on recursion. Thus, understanding how recursion works is imperative.

Next, I am going to show you the power of algorithms. We'll compare simple, unoptimized samples with ones which rely on algorithms. If you had any doubts about the importance of algorithms, these examples are going to convince you.

Then, we'll delve into generics. You must understand generics before we start studying data structures and algorithms.

Next, I'm going to walk you through the built-in Swift collection types.

Then we jump into the topic of basic sorting algorithms. We'll talk about selection sort, bubble sort, and insertion sort. We'll analyze the efficiency of each of these sorting algorithms, and we are going to represent them visually.

We'll then talk about two, more advanced sorting algorithms: merge sort and quicksort.

Finally, I'm going to share with you some useful online resources which will help you in sharpening your coding and problem solving skills.

After finishing this book, you'll know all the concepts related to algorithms and data structures. You'll have a good understanding of how the popular search algorithms work and the way they can be implemented using Swift.

As the first part of a series on Swift programming, this book is an essential steppingstone for delving deeper into algorithms, data structures and programming in general.

CHAPTER 2
THE BIG-O NOTATION

The *Big-O notation* is a mathematical model used in computer science to describe the efficiency of algorithms as a function of their input size.

The best way to understand the Big-O thoroughly is through code examples. Therefore, I'm going to illustrate each concept using Swift coding.

Here are the common orders of growth - or complexities - we are going to talk about in this chapter:

Constant Time - describes an algorithm that will always execute in the same amount of time, regardless of the input size.

Linear Time - describes an algorithm whose performance will grow linearly and in direct proportion to the size of the input data set.

Quadratic Time - represents an algorithm whose performance is directly proportional to the square of the size of the input data set. This behavior is typical with algorithms that involve nested iterations over the data set. Deeper nested iterations will result in O(N3) (cubic time), O(N4) quartic time and worse.

Logarithmic Time - represents a highly efficient algorithm, used by the binary search technique for example.

We'll cover the basics of Big-O. This knowledge is sufficient to understand the efficiency of the sorting algorithms presented in this course.

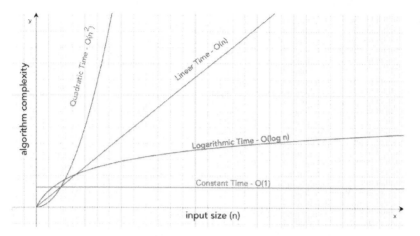

This graph visualizes the running times of some of the most popular sorting algorithms. As the input size increases, the performance differences become increasingly evident.

When the input count is small, all algorithms perform almost equally.

Actually, when testing with small datasets, we may even have the impression that the algorithm with a logarithmic complexity has the worst performance; however, as the size of the datasets grows, we'll clearly see the differences.

*If you want to follow along with me, download the repository from GitHub. The source code for this demo can be found in the **big-o-src** folder.*

SECTION 2.1
CONSTANT TIME COMPLEXITY

Constant Time describes an algorithm that will always require the same amount of time to execute, regardless of the input size.

Here's what we get if we visualize the running time of an algorithm that executes in constant time:

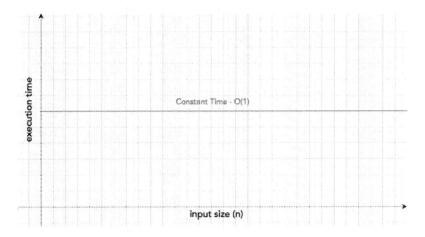

The input size doesn't affect the execution time. In other words, the execution time remains constant.

An algorithm which has a constant time complexity will run for the same amount of time whether it operates on one or on several thousand or million entries.

In the following demo, we are going to implement some examples that will execute in constant time.

Throughout this book, we're going to use Swift playgrounds.

We'll implement a utility class for measuring the performance. We'll use this utility class to perform measurements in the current demo and several other upcoming projects.

To illustrate the constant time complexity, we'll create a function that performs a check on an array.

The second example relies on the hash map lookup. We're going to compare the times required to retrieve values based on their keys from Swift dictionaries of various sizes. If implemented correctly, the hash map lookup should happen in constant time.

All right, now let's switch to Xcode.

> *If you want to follow along with me, download the repository from GitHub. Open the Big-O playground from the **big-o-src** folder. You can find the source code for this demo in the "Constant Time" playground page.*

First, we'll implement the benchmarking utility. The BenchTimer class has a single method called measureBlock(closure:).

```
import Foundation
import QuartzCore

public class BenchTimer {
    public static func measureBlock(closure:() -> Void) -> CFTimeInterval {
        let runCount = 10
        var executionTimes = Array<Double>(repeating: 0.0, count: runCount)
        for i in 0..<runCount {
            let startTime = CACurrentMediaTime()
            closure()
            let endTime = CACurrentMediaTime()
            let execTime = endTime - startTime
            executionTimes[i] = execTime
        }
        return (executionTimes.reduce(0, +)) / Double(runCount)
    }
}
```

The measureBlock(closure:) type method takes a closure argument. The measured code is executed ten times, and we store the individual run times in an array.

We rely on the QuartzCore framework's CACurrentMediaTime() function to retrieve the current absolute time.

Unlike NSDate or CFAbsoluteTimeGetCurrent(), CACurrentMediaTime() is reliable since it's not subject to changes in the external time reference.

The measured block gets executed between querying the startTime and endTime; the run time is then stored in the executionTimes array.

After ten iterations, we calculate the average execution time. The reduce() array function calls the closure sequentially on all the array elements, which in our case sums up all the items in the array.

Finally, we divide the result by the number of iterations, which gives us the average execution time.

Next, we implement a function, which takes an array and checks whether the first element is 0.

```
// Checks whether the first element of the array is zero
func startsWithZero(array: [Int]) -> Bool {
    guard array.count != 0 else {
        return false
    }
    return array.first == 0 ? true : false
}
```

This simple function executes in constant time: that is, the run time should be the same regardless of the size of the input array.

Let's prove our theory. We'll invoke the startsWithZero() function with three arrays. The first array has only three elements, the second one has 10,000 entries and the third array is huge, with 1,000,000 items.

```
var smallArray = [1,0,0]
var exectime = BenchTimer.measureBlock {
    _ = startsWithZero(array: smallArray)
}
print("Average startsWithZero() execution time for array with \(smallArray.count) elements:
```

```
\(exectime.formattedTime)")

var mediumArray = Array<Int>(repeating: 0, count: 10000)
exectime = BenchTimer.measureBlock {
    _ = startsWithZero(array: mediumArray)
}
print("Average startsWithZero() execution time for array
with \(mediumArray.count) elements:
\(exectime.formattedTime)")

var hugeArray = Array<Int>(repeating: 0, count: 1000000)
exectime = BenchTimer.measureBlock {
    _ = startsWithZero(array: hugeArray)
}
print("Average startsWithZero() execution time for array
with \(hugeArray.count) elements:
\(exectime.formattedTime)")
```

Our benchmark shows that the size of the input array does not affect the run time. There are only negligible differences in the order of microseconds.

Another algorithm which runs in constant time is the hash-map lookup. We will use the generateDict(size:) helper function to create custom-sized dictionaries.

```
let smallDictionary = ["one": 1, "two": 2, "three": 3]
exectime = BenchTimer.measureBlock {
    _ = smallDictionary["two"]
}
print("Average lookup time in a dictionary with
\(smallDictionary.count) elements:
\(exectime.formattedTime)")

// Generates dictionaries of given size
func generateDict(size: Int) -> Dictionary<String, Int> {
    var result = Dictionary<String, Int>()
    guard size > 0 else {
        return result
    }
    
    for i in 0..<size {
        let key = String(i)
        result[key] = i
    }
    return result
}

let mediumDict = generateDict(size: 500)
```

```
exectime = BenchTimer.measureBlock {
    _ = mediumDict["324"]
}
print("Average lookup time in a dictionary with
\(mediumDict.count) elements: \(exectime.formattedTime)")

let hugeDict = generateDict(size: 100000)
exectime = BenchTimer.measureBlock {
    _ = hugeDict["55555"]
}
print("Average lookup time in a dictionary with
\(hugeDict.count) elements: \(exectime.formattedTime)")
```

As you'll see after running the demo, the time it takes to find an element does not depend on the size of the dictionary. Constant time algorithms are great because they are not affected by any of the input parameters. However, it is not always possible to come up with a solution which runs in constant time.

SECTION 2.2
LINEAR TIME COMPLEXITY

Linear Time describes an algorithm whose performance will grow linearly and in direct proportion to the size of the input dataset.

This graph represents the running time of an algorithm which executes in linear time:

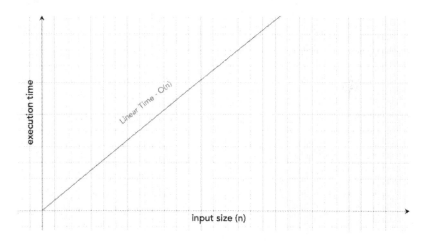

There is a 1-to-1 correlation between input size and execution time.

The run time of an algorithm with linear time complexity will increase at the same rate as the input dataset grows. For example, if one thousand entries took one second to process, then 10 thousand would require ten times as much, that is, 10 seconds. 100.000 entries would be processed in 100 seconds and so on. As the input dataset grows, so will our algorithms processing time increase, too. That's why it's called linear time.

Next, we'll implement some examples that will execute in linear time. First, I'm going to show you a function which sums up the elements of an array.

The next function we implement returns the number of odd and even integers from an array. Both functions will iterate through the entire array. Therefore the computation time is directly related to the size of the array. We have linear complexity in both cases. For performance measurements, we'll rely on the BenchTimer we built in the previous chapter.

> *If you want to follow along with me, download the repository from GitHub. Open the Big-O playground from the **big-o-src** folder. You can find the source code for this demo in the "Linear Time" playground page.*

To illustrate the linear time complexity, we'll need arrays of different sizes. So let's first implement a function which generates custom-sized arrays with random content.

```
// Generates random arrays of given size
func generateRandomArray(size: Int, maxValue: UInt32) -> [Int] {
    guard size > 0 else {
        return [Int]()
    }
    var result = [Int](repeating: 0, count: size)
    for i in 0..<size {
        result[i] = Int(arc4random_uniform(maxValue))
    }
    return result
}
```

The generateRandomArray(size:, maxValue:) function takes two arguments: the first gives the size of the array, and maxValue which determines the maximum allowed value for the elements. We use arc4random_uniform() to create random content for the array.

Next, we create a function which sums up the elements of the input array.

```
// Sums up the numbers from the input array
func sum(array: [Int]) -> Int {
    var result = 0
    for i in 0..<array.count {
        result += array[i]
    }
    return result
}
```

We could use the reduce() array function to calculate the sum, but implementing a custom solution makes it obvious that we iterate over the entire array.

We're going to test the sum(array:) function with three arrays of different sizes. The first array has 100 items, the next has 1000 and the last one has 10000 elements.

```
let array100 = generateRandomArray(size: 100, maxValue: UInt32.max)
var execTime = BenchTimer.measureBlock {
    _ = sum(array: array100)
}
print("Average sum() execution time for \(array100.count) elements: \(execTime.formattedTime)")

let array1000 = generateRandomArray(size: 1000, maxValue: UInt32.max)
execTime = BenchTimer.measureBlock {
    _ = sum(array: array1000)
}
print("Average sum() execution time for \(array1000.count) elements: \(execTime.formattedTime)")

let array10000 = generateRandomArray(size: 10000, maxValue: UInt32.max)
execTime = BenchTimer.measureBlock {
    _ = sum(array: array10000)
}
print("Average sum() execution time for \(array10000.count) elements: \(execTime.formattedTime)")
```

After executing our tests, the performance measurement values displayed in the console prove that the execution time is linear.

The sum(array:) function iterates through all the elements of the array. Thus, it is normal that the execution time increases proportionally with the size of the array.

Here's another function which needs to iterate through all the elements of the input list.

```
// Returns the number of odd and even elements from the input array
func countOddEven(array: [Int]) -> (even: UInt, odd: UInt)
```

```
{
    var even: UInt = 0
    var odd: UInt = 0

    for elem in array {
        if elem % 2 == 0 {
            even += 1
        } else {
            odd += 1
        }
    }
    return (even, odd)
}
```

The countOddEven(array:) function checks each item to find out whether it is odd or even.

Our tests confirm that countOddEven(array:) is indeed a function which runs in linear time - the execution time increases at the same rate as the input dataset grows.

SECTION 2.3
QUADRATIC TIME COMPLEXITY

Quadratic Time represents an algorithm whose performance is directly proportional to the square of the size of the input dataset.

As you can see in this graph, the runtime increases sharply, faster than the input sizes.

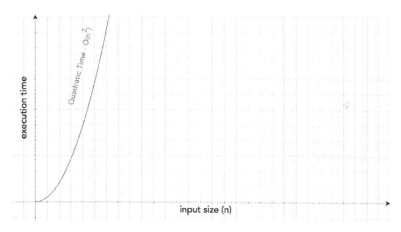

The runtime of an algorithm with quadratic time complexity will go up as a square of the input dataset size.

Quadratic, cubic and quartic time complexities are a result of nested operations on a dataset. You should try to avoid them whenever possible, due to the negative impact on the overall performance.

Let's say your quadratic-time algorithm processes 100 entries in 100 milliseconds. For 2000 entries the algorithm would run for 40 seconds. And 4000 entries would be processed in slightly more than 26 minutes!

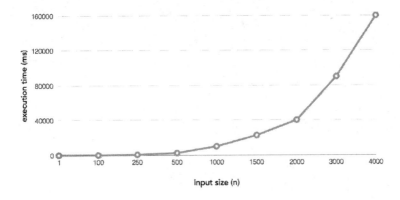

The runtime grows even more sharply with the input size in the case of cubic or quartic time complexity.

In the following demo, we are going to build a function that creates multiplication tables. The function will use two nested loops; because of the nested iterations, this algorithm has a quadratic time complexity.

Now let's switch to XCode.

> *If you want to follow along with me, download the repository from GitHub. Open the Big-O playground from the **big-o-src** folder. You can find the source code for this demo in the "Quadratic Time" playground page.*

First, we implement a useful addition for our benchmarking utility.

```
public extension CFTimeInterval {
    public var formattedTime: String {
        return self >= 1000 ? String(Int(self)) + "s"
            : self >= 1 ? String(format: "%.3gs", self)
            : self >= 1e-3 ? String(format: "%.3gms", self * 1e3)
            : self >= 1e-6 ? String(format: "%.3gµs", self * 1e6)
            : self < 1e-9 ? "0s"
            : String(format: "%.3gns", self * 1e9)
    }
}
```

We add a formattedTime property to the CFTimeInterval type. This property provides a concise string representation of the time interval value which also includes the right unit of time, which ranges from nanoseconds to seconds.

Next, we'll implement a function to demonstrate the quadratic time complexity.

```
// Generates multiplication tables
func multTable(size: Int) -> [Int] {
    var table = [Int]()
    let array = [Int](1...size)

    for i in 0..<array.count {
        for j in 0..<array.count {
            let val = array[i]*array[j]
            table.append(val)
        }
    }
    return table
}
```

The multTable(size:) function takes an integer argument which gives the size of the array which holds positive integers in the range 1..size. The function returns the results of multiplying each element in the array with every other value.

We use two loops to compute the result: the outer loop iterates through the indices of the array. The internal loop takes the value found at the outer index and multiplies it with every item from the same array. The output is a multiplication table - let's check out what we get for the input value 10.

Now let's analyze how the two nested loops influence the processing time.

For a two-element array, the outer loop iterates two times, and the internal loop also two times for every outer iteration. This gives us four iterations in total. For a three element array, the total iteration count is 3 times 3, that is, 9. For ten elements the function will loop 100 times.

The number or iterations goes up as a square of the input data size.

Let's run some performance tests to prove that our function runs at quadratic time complexity. We call the multTable(size:) function with arrays of different sizes, and we'll display the run times in the console.

```
let sizes = [10, 20, 30]

for i in 0..<sizes.count {
    let size = sizes[i]
    let execTime = BenchTimer.measureBlock {
        _ = multTable(size: size)
    }
    print("Average multTable() execution time for \(size) elements: \(execTime.formattedTime)")
}
```

After executing the demo, we can see the quadratic jumps in execution time.

It is worth mentioning that - especially for smaller input - the measurements might not always reflect the quadratic time complexity because of under the hood compiler and hardware optimizations.

SECTION 2.4
HINTS FOR POLYNOMIAL TIME COMPLEXITY

Nested loops within a method or function should always be a warning sign, and you should avoid them at all costs.

Whenever you encounter two or more loops that resemble Russian nesting dolls, ask yourself whether the nested iterations are really required to solve that particular problem. First, document the questionable code using dedicated comments (!!! or XXX) - or even better, add a *warning* if supported by the given compiler. Then, implement unit tests to highlight the performance issue caused by the nested loops. Finally, try to solve the problem by replacing the nested iterations with a more efficient algorithm.

Polynomial time complexity is usually the result of rushed coding, produced by developers who lack time or expertise - or maybe both.

More often than not, you'll find a better solution.

> *In the chapter called "The Power of Algorithms", we'll see examples of replacing inefficient approaches with solutions that bring huge performance gains.*

There may be cases when there really is no better way to solve that problem, and you can't get rid of the nested loops. Then document the affected method thoroughly and describe why it works that way. Also, describe the performance issues it may cause with larger datasets.

SECTION 2.5
LOGARITHMIC TIME

Logarithmic Time represents an extremely efficient algorithm, used by advanced algorithms like the binary search technique.

Logarithmic time means that time goes up linearly while the input data size goes up exponentially.

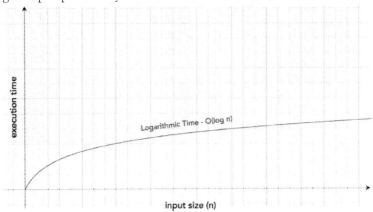

For example, if it takes 1 millisecond to compute 10 elements, it will take 2 milliseconds to compute 100 elements, 3 milliseconds to compute 1000 elements, and so on.

Binary search, quick sort and divide and conquer type of algorithms run usually in logarithmic time.

First, let's take a closer look at the logarithm.

In mathematics, the logarithm is the inverse operation to exponentiation. The logarithm of x to base b: $log_b(x) = y$ is the unique real number y such that: $b^y = x$

For example, $log_{10}(1000) = 3$, as $10^3 = 1000$. And $log_4(16) = 2$, because $4^2 = 16$

In other words,

> *The logarithm of a number is the exponent to which another fixed number, the base, must be raised to produce that number.*

$$\log_b(x) = y$$
$$b^y = x$$

In computer science, when measuring the performance of algorithms, the base of the logarithm is not specified, because the result only changes by a constant factor when another base is used. A constant factor is usually disregarded in the analysis of algorithms.

SECTION 2.6
SUMMARY

We dedicated this entire chapter to the Big-O notation.

Understanding time complexity paves the road to working with algorithms. We've talked about constant time complexity - where the execution time is constant and does not depend on the input size.

Checking the first element of an array or retrieving an item from a dictionary are good examples for the constant time complexity.

Linear time complexity describes an algorithm whose runtime grows in direct proportion to the size of the input. For example, enumerating through the elements of an array works in linear time.

The execution times of quadratic time algorithms go up as a square of the input dataset size. Quadratic time complexity is produced by a loop nested into another loop, as we've seen in our multiplication table example.

Try to avoid polynomial time complexity - like quadratic, quartic or cubic - as it can become a huge performance bottleneck.

Logarithmic time describes complex algorithms like the quicksort and shows its benefits when working with larger data sets.

CHAPTER 3
RECURSION

In programming, repetition can be described using loops, such as the for-loop or the while loop. Another way is to use recursion.

We encounter recursion frequently while studying algorithms and data structures. Thus, it is important to understand what recursion is.

I'm going to show you how recursion works through live coding examples.

Recursion is a useful technique, yet it doesn't come without pitfalls. We'll finish this chapter by demonstrating how to avoid common issues when using recursion in Swift projects.

SECTION 3.1
WHAT'S RECURSION?

By definition, recursion is a way to solve a reoccurring problem by repeatedly solving similar subproblems.

In programming, we can have recursive functions. A function is recursive if it calls itself. The call can happen directly like in this case:

```
func r() {
    //...
        r()
    //...
}
```

Or indirectly, if the function calls another function, which in turn invokes the first one:
```
func r() {
    //...
        g()
    //...
}

func g() {
    //...
    r()
    //...
}
```

We'll often encounter recursive data structures, too. A data structure is recursive if you can describe it in terms of itself.

A linked list can be described as a list node followed by a linked list.

Here's a simple Node class:

```
class Node {
    var next: Node?
    var value: String

    init(value: String) {
        self.value = value
    }
}
```

Each Node can link to the next node through the next property.

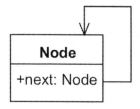

The Node also holds a value of type string. I provide an initializer which sets the value property.

Now that we have our Node type, let's build a linked list.

```
let node1 = Node(value: "node1")
let node2 = Node(value: "node2")
let node3 = Node(value: "node3")node1.next = node2
node2.next = node3
node3.next = nil
```

First, I create each node using a value that matches its name. Then, I assign the next property of each node to form a singly linked list. I end the linked list after three nodes by setting the next property to nil.

Let's implement a function that traverses the linked list and prints the value in each node.

```
func parseNodes(from node: Node?) {
    guard let validNode = node else {
        return
    }
    print(validNode.value)
    parseNodes(from: validNode.next)
}
```

The parseNodes(from:) function takes an argument of type Node. This is the node where we start the list traversal.If the node is nil, we exit the function. Else, we print the value of the given node.The function calls the parseNodes(from:) method recursively by passing in the next node.
parseNodes(from: node1)

Finally, we call the parseNodes(from:) function with the first node as input

parameter.
If we run the demo, it prints the expected values. Since the data structure is *recursive*, we can use recursion to iterate through it.

Recursion won't necessarily produce faster or more efficient code. But it usually provides an elegant alternative to iterative approaches and requires fewer lines of code.

SECTION 3.2
HOW DOES RECURSION WORK?

So far, we've seen some examples of recursive functions and data structures. Now, let's check out how recursion works.

We're going to calculate the factorial of a positive integer n. This is a problem that can be solved using recursion, since the factorial is defined as the product of the integers from 1 to n.

$n! = 1 \times 2 \times 3 \times \ldots \times n$

So, here's the Swift factorial function that calculates the factorial of a positive integer:

```
func factorial(n: UInt64) -> UInt64 {
    return n < 2 ? 1 : n * factorial(n: n - 1)
}
```

The function takes an unsigned integer as argument. If the input parameter is 1 or 0, the function returns 1. Otherwise, the function returns the product of the input and the result of calling the function with an argument that is smaller by one. The recursive calls continue until the function gets called with a value that is smaller than 2.

To understand how recursion works, here's a graphical representation of what's going on when calculating factorial of 3.

Whenever a nested call happens, the execution of the former call is suspended and its state is stored. A snapshot of its context, that is, its code, input parameters and local variables is persisted.

All this information is stored in a structure known as "call stack" or "execution stack." The call stack is a stack structure that keeps track of the point where control should be returned after a subroutine finishes its execution.

When the nested call is finished executing, its context is destroyed, and the control is returned to the caller.

Eventually, we get back to the very first function call. All the nested contexts are destroyed by then and the result is returned to the caller.

SECTION 3.3
RECURSION PITFALLS

Recursion is great, but it doesn't come without pitfalls.

The biggest problem is infinite recursion. I'm going to illustrate it using a function which calculates the sum of the first n positive integers.

> *I call the function badSum(), to make it clear that it's not the right approach to solve this problem. Calculate Sum(n) shows the proper solution that relies on a simple formula.*

```
func badSum(n: Int) -> Int {
    return n + badSum(n: (n - 1))
}
```

So, badSum(n: Int) accepts an input parameter n, of type Int, and returns an integer. We use a recursive approach. The function returns the sum of the input parameter and the result of calling itself with an argument that is smaller by one.

We can test this function in an Xcode playground like this:

```
let res = badSum(n: 3)
print(res)
```

Eventually, our demo is going to crash.

To understand the root cause, let's quickly recap how recursion works.

Each time a nested call occurs, a record of the current context is made and added as a new stack frame to the top of the stack.

And each time a call returns, the top stack frame is removed.

The stack is a special part of the memory used for fast, static allocations. Each application has a stack (and multithreaded applications have one stack per thread).

The stack has a finite size. Thus, if we keep calling nested functions and none of these functions returns, we'll run out of available stack space.

When the memory available for the call stack is exceeded, the app will crash with a stack overflow error.

Let's inspect our badSum(n:) function.

```
func badSum(n: Int) -> Int {
    return n + badSum(n: (n - 1))
}
```

There's no condition that prevents executing the nested call over and over again.

To see what's going on, we print the value of the input argument. So, whenever we call the function, the value of n will appear in the console.

```
func badSum(n: Int) -> Int {
    print(n)
    return n + badSum(n: (n - 1))
}
```

If we run the demo after this change, negative values will start filling the console. Now, we have the proof that the execution doesn't stop after two recursive calls as it should.

That means that the nested contexts are not destroyed. Hence, the stack frames are not removed. Thus, we keep allocating memory on the stack without deallocating it, which eventually leads to the stack overflow exception.

Uncontrolled recursion leads to stack overflow!

SECTION 3.4
HOW TO AVOID INFINITE RECURSION?

To avoid infinite recursion, each recursive function needs to have at least one *condition that prevents further nested calls* to itself. This condition is called base case.

The issue is that badSum(n:) calls itself as we keep decreasing the input argument n.

```
func badSum(n: Int) -> Int {
    return n + badSum(n: (n - 1))
}
```

We need a base case here - that is, a condition which makes the function return without performing another recursive call.
Let's fix the badSum(n:) function and add the missing base case. We're only interested in the sum of positive integers. So, if the input argument is zero, we can simply return zero.

```
func badSum(n: Int) -> Int {

    if n == 0 {
        return 0
    }

    return n + badSum(n: (n - 1))
}
```

If I run the function after this change, it produces the expected result.

But does this check really work for all input values? Will it work for negative values? Let's check it with -1.

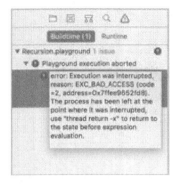

Nope! Since I only check for zero, the function will cause a runtime crash for negative input.

We must ensure that the function actually *progresses towards the base case*.

For that, we need to modify the base case so that it covers not only the value zero, but also negative values.

```
func badSum(n: Int) -> Int {

    if n == 0 {
    if n <= 0 {
        return 0
    }

    return n + badSum(n: (n - 1))
}
```

Actually, the guard statement is an even better choice, since it forces us to exit the function:
```
func badSum(n: Int) -> Int {

    if n == 0 {
    guard n > 0 else {
        return 0
    }

    return n + badSum(n: (n - 1))
}
```

> *Some programming languages have built-in checks to prevent stack overflows.*

Remember these rules when you implement recursive solutions:

- Each recursive function needs to have at least one *condition that prevents further nested calls* to itself

- Ensure that the function actually *progresses towards the base case*
  Also, check your recursive function thoroughly through unit tests that cover also edge cases.

CHAPTER 4
THE POWER OF ALGORITHMS

In this chapter, we're going to take a closer look at the importance and the benefits of algorithms and algorithmic thinking.

We've already talked about the Big-O notation. We saw that our choice of implementing a problem can make a huge difference when it comes to the performance and the scalability of our solution.

Algorithmic thinking is the ability to approach a problem and find the most efficient technique to solve it.

To demonstrate the power of algorithms, we are going to solve the following problems:

- We'll calculate the sum of the first N natural numbers.

- Then, we are going to implement a function that, given an array and a value, returns the indices of any two distinct elements whose sum is equal to the target value

- Finally, we are going to calculate the equilibrium index of an array. The equilibrium or balance index represents the index which splits the array such that the sum of elements at lower indices is equal to the sum of items at higher indices.

We're going to use a brute-force approach first. Then we'll implement a solution with efficiency in mind.

You'll see how some basic math and the knowledge of Swift language features and data structures can help us in implementing a cleaner and better performing solution.

SECTION 4.1
CALCULATE SUM(N)

Our first task is to implement a function which calculates the sum of the first N natural numbers.

We'll start with a naive implementation. Then, we are going to implement a more efficient way to solve this problem using a formula that is more than 2000 years old.

All right, so let's switch to our Xcode playground project.

> *If you want to follow along with me, download the repository from GitHub. Open the Sum(N) playground from the **algorithm-power-src** folder.*

```
func sum(_ n: UInt) -> UInt {
    var result: UInt = 0
    for i in 1...n {
        result += i
    }
    return result
}
```

I define a function called sum() which takes a single argument; this argument represents the number of natural numbers whose sum we want to calculate. The function returns an unsigned integer which gives the sum. The logic is very simple: we sum up all the numbers starting with one up to the value of the input parameter n.

```
for i in 1...n {
    result += i
}
```

Because of this loop, our function has a linear time complexity - O(n). We can confirm the linear time complexity by calling the sum() function with steadily increasing values for the step constant (e.g. 100, 1000, 10000).
let step = 100

var execTime: Double

```
for i in 1...10 {
    execTime = BenchTimer.measureBlock {
        _ = sum(UInt(i*step))
    }
    print("Average sum(n) execution time for \(i*step) elements: \(execTime.formattedTime)")
}
```

The console log shows that the execution time increases linearly with the input.

Although this function does the job, there is a better way to compute the sum of the first N natural numbers. We are going to rely on a simple formula:

$$\sum_{k=1}^{n} k = \frac{n \times (n + 1)}{2}$$

Carl Friedrich Gauss is said to have found this relationship in his early youth. However, he was not the first to discover this formula. It is likely that its origin goes back to the Pythagoreans and it was known as early as the sixth century BC.

We can now implement an improved version of the sum() function.

```
func sumOptimized(_ n: UInt) -> UInt {
    return n * (n + 1) / 2
}
```

sumOptimized() does not rely on loops. Instead, it uses the triangle numbers formula.

The new function is not only cleaner, but it also operates in constant time; that is, its execution time doesn't depend on the input.

You can check this by running the same performance tests as we did for the sum() function. The results will prove that the execution times do not vary regardless of the input size. There will be only some minor, negligible differences in the range of μs.

The sumOptimized is more efficient even for smaller values, and the difference just grows with the input. This chart visualizes the running times of the two functions:

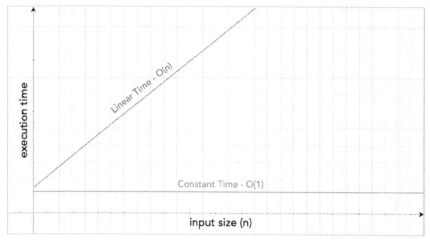

The optimized, sumOptimized() function doesn't depend on the input size, unlike the sum() function which runs in linear time.

By applying this clever formula, we managed to implement a solution with an optimal performance.

SECTION 4.2
PAIR MATCHING CHALLENGE

Here's the next challenge. Our task is to write a function that, given an array and a target value, returns zero-based indices of any two distinct elements whose sum is equal to the target sum. If there are no such elements, the function should return nil.

For example, for the array of numbers 1, 2, 2, 3, 4 and the target value 4, our function should return the tuple of indices (1,2) or (0,3).

Problem

```
  0  1  2  3  4
[1, 2, 2, 3, 4]
Target = 4
```

Solution

(1,2)   2 + 2 = **4**
(0,3)   1 + 3 = **4**

First, we are going to implement a solution which relies on nested iterations. Because of the nested iterations, this algorithm has a quadratic time complexity. Then, we are going to come up with an algorithm which operates in linear time rather than quadratic time.

All right, time to do some coding!

> *If you want to follow along with me, download the repository from GitHub. Open the FindTwoSum playground from the **algorithm-power-src** folder.*

INTRODUCTION TO ALGORITHMS AND DATA STRUCTURES IN SWIFT 4

Here's the first attempt to solve this problem:

```swift
func findTwoSum(_ array: [Int], target: Int) -> (Int, Int)?
{
    guard array.count > 1 else {
        return nil
    }

    for i in 0..<array.count {
        let left = array[i]
        for j in (i + 1)..<array.count {
            let right = array[j]
            if left + right == target {
                return (i, j)
            }
        }
    }
    return nil
}
```

We step through the array and check every value. Starting from the outer index, we scan the rest of the array. If we find a number such that the two numbers add up to the target, we return the tuple. Else, we continue iterating through the subarray.

This algorithm uses two nested loops, which means quadratic time complexity. To be precise, this function has a worst time complexity of $\frac{n^2 + n}{2}$, where n is the size of the input array.

Let's go through the steps of finding this formula. The outer loop iterates through the entire array, which gives n iterations. The inner loop iterates (n - 1) times first, then (n - 2) times, and so on.

Outer loops   Inner loops
n   (n - 1) + (n - 2) + ... + 1

When the outer loop reaches the penultimate index, the inner loop only iterates once. To find out the total iterations performed in the inner loop, we have to calculate the sum of the first (n - 1) numbers.

We can use the formula we learned in the previous lecture to calculate the inner loop count, that is:

Inner loops
$$(n-1) + (n-2) + \ldots + 1 = \frac{(n-1) \times (n-1+1)}{2} = \frac{(n-1) \times n}{2}$$

Now, to get the total number of iterations, we must add the count of the outer loops, too:

$$\text{Total} = \text{Outer loops} + \text{Inner loop}$$

$$n + \frac{(n-1) \times n}{2} = \frac{2n + n^2 - n}{2} = \frac{n^2 + n}{2}$$

So, our function has a time complexity of $\frac{n^2+n}{2}$

Now, let's implement a solution which doesn't rely on nested loops. We're going to avoid scanning the array twice to find numbers which add up to the target value.

```
func findTwoSumOptimized(_ array: [Int], target: Int) -> (Int, Int)? {
    guard array.count > 1 else {
        return nil
    }
    
    var diffs = Dictionary<Int, Int>()
    
    for i in 0..<array.count {
        let left = array[i]
        
        let right = target - left
        
        if let foundIndex = diffs[right] {
            return (i, foundIndex)
        } else {
            diffs[left] = i
        }
    }
    
    return nil
}
```

findTwoSumOptimized(_, target:) uses a single loop to iterate through the array. For each number, we check whether the difference between the target value and the given number can be found in the dictionary called *diffs*.

If the difference is found, we've got our two numbers, and we return the tuple with the indices. Else, we store the current index (the difference being the key) and we iterate further.

> *Note that both dictionary insertion and search happen in constant time. Therefore, these operations won't affect the time complexity of our function.*

We managed to reduce the time complexity to O(n) - linear time - because we only iterate once through the array.

If we compare the performance of the two functions, we'll see noticeable differences as the size of the input array increases.

The benefits of the linear complexity over quadratic time complexity become even more evident if we increase the input size further.

SECTION 4.3
FIND THE EQUILIBRIUM INDEX

The equilibrium index of an array is an index such that the sum of elements at lower indices is equal to the sum of elements at higher indices.

For example, in an array A with the numbers [-3, 2, -2, 1, -2], 1 and 2 are equilibrium indices:

Problem $A = [\underset{0}{-3}, \underset{1}{2}, \underset{2}{-2}, \underset{3}{1}, \underset{4}{-2}]$

Solution
1. $A[0] = A[2] + A[3] + A[4]$
2. $A[0] + A[1] = A[3] + A[4]$

1 is an equilibrium index, because: $A[0] = A[2] + A[3] + A[4]$

that is, $-3 = -2 + 1 + (-2)$

2 is also an equilibrium index, because the sum of elements at indices 0 and 1 is equal to the sum of elements at indices 3 and 4: $-3 + 2 = 1 + (-2)$

We're going to implement a function that given an array returns the equilibrium indices or nil if the array has no equilibrium index.

First, we'll come up with a brute-force solution. Then, we'll come up with an algorithm that executes in linear time by applying some basic math.

> *If you want to follow along with me, download the repository from GitHub. Open the EquilibriumIndex playground from the* **algorithm-power-src** *folder.*

```swift
func equilibrium(_ numbers: [Int]) -> [Int]? {
    guard numbers.count > 1 else {
        return nil
    }
    
    var indices = [Int]()
    
    var left = 0
    var right = 0
    
    let count = numbers.count
    
    for i in 0..<count {
        left = 0
        right = 0
        
        for j in 0..<i {
            left = left + numbers[j]
        }
        
        for j in i+1..<count {
            right = right + numbers[j]
        }
        
        if left == right {
            indices.append(i)
        }
    }
    
    return indices.isEmpty ? nil : indices
}
```

The equilibrium() function uses three loops. The outer loop iterates through all the elements of the array.

We need two inner loops to find out whether the current index picked by the outer loop is an equilibrium index or not. The first inner loop tracks the sum of the elements at lower indices. The second inner loop keeps track of the sum of items at higher indices.

In other words, we split the array at the position of the outer index. Then, we compare the sum of the elements of the two subarrays. If the sums are equal, we found an equilibrium index, and we add it to the array of indices. Else we keep iterating further.

The worst time complexity of this solution is quadratic. Updating the left and right sums using two inner loops is a simple but also a very inefficient approach.

Here's a solution that doesn't require nested loops.

```swift
func equilibriumOptimized(_ numbers: [Int]) -> [Int]? {
    var indices = [Int]()
    
    var leftSum = 0
    var sum = numbers.reduce(0, +)
    let count = numbers.count
    
    for i in 0..<count {
        sum = sum - numbers[i]
        
        if leftSum == sum {
            indices.append(i)
        }
        
        leftSum = leftSum + numbers[i]
    }
    
    return indices.isEmpty ? nil : indices
}
```

The idea is to get the total sum of the array first. We use the reduce(0, +) array function to calculate the sum of all the elements in the array.

Then, we iterate through the entire array. The leftSum variable is originally initialized to zero. We keep updating leftSum by adding the elements as we iterate through the array.

We can get the sum of the elements of the right subarray by subtracting the elements one by one from the total sum.

With this algorithm, we only need to loop once through the array. This function has a linear time complexity, $O(2n)$ because of the reduce() function and the loop.

Let's run our performance tests. For 5 items, the optimized variant performs about 2x better, and over 20x times better for 50 items. For an array of 200 items, the equilibriumOptimized() function runs about 1000x faster than the function which uses nested loops.

The difference grows quickly with the input size. For bigger arrays, finding the equilibrium indices with the optimized variant only takes a fraction of the running time of the brute-force approach.

As we've expected, the function with quadratic time complexity is more sensitive to input data sizes than the function with linear time complexity.

This is another good example which demonstrates the importance of finding the right algorithm.

SECTION 4.4
SUMMARY

In this section, we've seen some practical examples of solving problems using two different approaches.

Although the naive implementations produced the right results, they start to show their weaknesses as the input size gets bigger.

By using more efficient techniques, we reduced the time complexity and - as a consequence - the execution time of our solutions considerably.

Coming up with the optimal algorithm requires research and deeper understanding of the problem we are trying to solve. Math skills and the ability to apply the features of the given programming language will help you in creating more efficient algorithms.

The time complexity of an algorithm is crucial when it comes to performance.

Do your best to avoid polynomial and worse time complexities!

Nested loops should always be a warning sign. Whenever you meet them, think about other alternatives to solve that particular problem.

You should highlight the performance bottlenecks using dedicated comments. Implement unit and performance tests. They will surface the issues which otherwise would remain hidden.

Finally, try to solve the problem without relying on nested iterations if possible.

CHAPTER 5
GENERICS

Generics stand at the core of the Swift standard library. They are so deeply rooted in the language that you can't avoid them. In most cases, you won't even notice that you're using generics.

SECTION 5.1
WHY GENERICS?

To illustrate the usefulness of generics, we'll try to solve a simple problem.

> *If you want to follow along with me, download the repository from GitHub. Open the Generics playground from the **generics-src** folder. You can find the source code for this demo in the "Pair without Generics" playground page.*

Let's say that we need to create types that hold pairs of different values.

Assuming that we need a pair type that contains two Strings, we could create a structure like this:

```
// Pair holding two String values
struct StringPair {
    var first: String
    var second: String
}
```

Next, we need a pair holding two Int values. So, we declare our IntPair type:

```
struct IntPair {
    var first: Int
    var second: Int
}
```

Then, we'll also need a type which has two floats:

```
struct FloatPair {
    var first: Float
    var second: Float
}
```

How about a Pair which has two Data properties? We'll create yet another pair structure called DataPair.

```
struct DataPair {
    var first: Data
    var second: Data
```

}

And if we need a String-Double pair? We create the appropriate type, right? Could it be simpler?

```
struct StringDoublePair {
    var first: String
    var second: Double
}
```

We can now use our newly created types whenever we need them. We must only remember their names. Whenever we need a new pair type, we simply create it.

```
struct StringPair {              struct IntPair {              struct FloatPair {
    var first: String                var first: Int                var first: Float
    var second: String               var second: Int               var second: Float
}                                }                             }

struct DataPair {                struct StringDoublePair {     struct StringDataPair {
    var first: Data                  var first: String             var first: String
    var second: Data                 var second: Double            var second: Data
}                                }                             }

struct IntDataPair {             struct FloatDataPair {        struct StringDatePair {
    var first: Int                   var first: Float              var first: String
    var second: Data                 var second: Data              var second: Date
}                                }                             }
```

Is this really the way to go? Definitely not!
We must stop the type explosion before it gets out of hand!

SECTION 5.2
GENERIC TYPES

Wouldn't it be cool to have only one type which can work with any value? Generic types come to the rescue!

> *If you want to follow along with me, download the repository from GitHub. Open the Generics playground from the **generics-src** folder. You can find the source code for this demo in the "Generic Types" playground page.*

Using generics, we can define a Pair structure that can be used with any type.

```
struct Pair<T1, T2> {
    var first: T1
    var second: T2
}
```

We use placeholders rather than concrete types for the properties in the structure. T1 and T2 are placeholders that can be used anywhere within the type's definition.

Also, we can simply call our struct Pair. Since it's a generic type, we don't have to specify the supported types in the name of the structure.

Now we can create pairs of any type using the following syntax:

```
// Float - Float
let floatFloatPair = Pair<Float, Float>(first: 0.3, second: 0.5)
```

We can even skip the placeholder types - the compiler is smart enough* to figure out the type based on the arguments.

```
// String - String
let stringAndString = Pair(first: "First String", second: "Second String")
// String - Double
```

```
let stringAndDouble = Pair(first: "I'm a String", second:
99.99)
// Int - Date
let intAndDate = Pair(first: 42, second: Date())
```

Type inference works by examining the provided values while compiling the code. In Swift, we can define our generic classes, structures or enumerations just as easily.

Generic types helped us stop the type explosion. Our code not only became shorter, but it's also reusable.

Next, we're going to talk about generic functions.

SECTION 5.3
GENERIC FUNCTIONS

Generic functions are another powerful feature of the Swift language.

A generic function or method can work with any type. Thus, we can avoid duplications and write cleaner code.

Let's start with a programming challenge: we need to implement a method which tells whether two values are equal.

> *If you want to follow along with me, download the repository from GitHub. Open the Generics playground from the **generics-src** folder. You can find the source code for this demo in the "isEqual without Generics" playground page.*

First, we define the isEqual function for Strings:

```
func isEqual(left: String, right: String) -> Bool {
    return left == right
}
```

Then, we need the same feature for Double types. Not a big deal.
```
func isEqual(left: Double, right: Double) -> Bool {
    return left == right
}
```

What about dates? Luckily, that won't take too long to implement either.
```
func isEqual(left: Date, right: Date) -> Bool {
    return left == right
}
```
Then, we also need to tell whether two Data instances are equal.

```
func isEqual(left: Data, right: Data) -> Bool {
    return left == right
}
```

Done!
By now, you probably see where this goes.

```
func isEqual(left: String, right: String) -> Bool {
    return left == right
}
func isEqual(left: Double, right: Double) -> Bool {
    return left == right
}
func isEqual(left: Date, right: Date) -> Bool {
    return left == right
}
func isEqual(left: Data, right: Data) -> Bool {
    return left == right
}
```

This is not the way to go!

Implementing a new function for every new type leads to a lot of redundant code. Such a code-base is hard to maintain and use.

We should always avoid code repetition. And generics help us solve this problem, too.

Let's create the generic isEqual function.

> *If you want to follow along with me, download the repository from GitHub. Open the Generics playground from the **generics-src** folder. You can find the source code for this demo in the "Generic Functions" playground page.*

The syntax for writing a generic function is similar to what we used to declare generic types.

```
func isEqual<T> (left: T, right: T) -> Bool {
    return left == right
}
```

We specify the placeholder type between angle brackets after the name of the function or method. We can refer to this placeholder type in the argument list or anywhere in the function's body.

If we compile our code as it is now, we get an error. This is because the compiler doesn't know how to compare two instances of the placeholder type. We must ensure that the type implements the equal-to operator. Types that adopt the Equatable protocol must implement the equal-to operator. So, we must enforce a type constraint. Type constraints specify that a type must conform to a particular protocol or inherit from a specific class. Let's add the Equatable type constraint.

```
func isEqual<T: Equatable> (left: T, right: T) -> Bool {
    return left == right
}
```

This limits the types that can be used as arguments in our function. We got rid of the compiler error.
Now, the generic isEqual function can only accept instances of types that adopt the Equatable protocol. Contact is a structure that doesn't conform to the Equatable protocol:

```
struct Contact {
    let name: String
    let address: String

    init(_ name: String, address: String) {
        self.name = name
        self.address = address
    }
}

let oldCampus = Contact("Old Apple Campus", address: "1 Infinite Loop, Cupertino, CA 95014")
let newCampus = Contact("New Apple Campus", address: "19111 Pruneridge Ave, Cupertino, CA 95014")

print(isEqual(left: oldCampus, right: newCampus))
```

Our code won't compile if we try to use the isEqual function with two Contact instances. So, let's add the Equatable protocol conformance.
```
struct Contact: Equatable {
    //...
}
```

To adopt the Equatable protocol, we have to implement the equal-to operator as a static member.

```
struct Contact: Equatable {
    let name: String
    let address: String
```

```swift
init(_ name: String, address: String) {
    self.name = name
    self.address = address
}

static func == (lhs: Contact, rhs: Contact) -> Bool {
    return lhs.name == rhs.name && lhs.address == rhs.address
}
}
```

The implementation is simple. We check whether the properties of the two arguments match.

The Equatable protocol also defines the not-equal operator. We don't need to implement it, though. The standard library provides a default implementation. This calls the custom equal-to operator and negates the result.

We can apply type constraints also to generic types. You may recall the generic Pair struct that we implemented in the previous lesson. Let's assume that we want to limit the types to those that implement the Comparable protocol.

```swift
struct Pair<T1: Comparable, T2: Comparable> {
    var first: T1
    var second: T2
}
```

With this type constraint, we can only create Pair instances out of types that adopt the Comparable protocol.

Generics are super useful. And Swift makes it easy for us to implement and use generic types and functions.

CHAPTER 6
THE BUILT-IN SWIFT COLLECTION TYPES

In this section, we're going to take a closer look at the built-in Swift collection types.

Swift provides three primary collections. We'll talk about:

- The Array - which is an ordered sequence of items

- The Set - which is an unordered sequence of unique values,

- And the Dictionary, that lets us store and access key-value pairs.

In Swift, the Array, the Set, and the Dictionary are implemented as generic types. As We've seen in the chapter about Generics, this provides great flexibility, as we can use any type with these generic collections. They can store instances of classes, structs, enumerations, or any built-in type like Int, Float, Double, etc.

SECTION 6.1
THE ARRAY

Arrays store values of the same type in a specific order. The values must not be unique: each value can appear multiple times. In other words, we could define the Array as an ordered sequence of non-unique elements.

> *If you want to follow along with me, download the repository from GitHub. Open the SwiftCollectionTypes playground from the* **collections-src** *folder. You can find the source code for this demo in the playground page "The Array".*

We create an array using the following syntax:

```
let numbers: Array<Int> = [1, 2, 5, 3, 1, 2]
```

Since it's a generic type, we must provide the type that our array can store. This is an array of Int types. What that means, is that we can use our array only with Ints.
There is a shorter way to define an array by placing the type between angle brackets.

```
let numbers: [Int] = [1, 2, 5, 3, 1, 2]
```

This is the *shorthand form*, and it is the preferred way of defining an array. Actually, there's an even shorter way to create an array of Ints.

```
let numbers = [1, 2, 5, 3, 1, 2]
```

For certain types, we can rely on Swift's type inference to work out the type of the array. Now, let's talk a bit about type inference.

Swift can guess the type based on the value we provide, which can save us from a lot of typing. However, type inference has its limitations.

For example, if we defined our array like this, the type inference engine would assume that it's an array of Double:

```
let array = [1.0, 2.0, 5.0]
```

To define an array of floats, we must explicitly provide the Float type.

```
let floats: [Float] = [1.0, 2.0, 5.0]
```

Each item in an array has an index associated with it. The indices start at zero, and their value gets incremented for each subsequent item. So, for our original example, the indices look like this:

let numbers = [1, 2, 5, 3, 1, 2]

Index	0	1	2	3	4	5
Value	1	2	5	3	1	2

We can iterate through the array and print the indices using the Array index(of:) instance method.

```
for value in numbers {
    if let index = numbers.index(of: value) {
        print("Index of \(value) is \(index)")
    }
}
```

After running this code in Xcode, the console log shows the following:
Index of 1 is 0
Index of 2 is 1
Index of 5 is 2
Index of 3 is 3
Index of 1 is 0
Index of 2 is 1
Note that index(of:) returns the first index where the specified value appears in the array. That's why we get the same index for the values that appear twice (1 and 2).

Another way is to use the array's forEach() method. This method executes its closure on each element in the same order as a for-in loop.

```
numbers.forEach { value in
```

```
    if let index = numbers.index(of: value) {
        print("Index of \(value) is \(index)")
    }
}
```

And the result is the same as with the for-in example.

SECTION 6.2
ACCESSING THE ARRAY

We can access the elements of an array by index.
The array has two convenience accessors for the first and the last element.

```
let first = numbers.first
let last = numbers.last
```

Note that first and last return optional values. If the array is empty, their value will be nil.
We don't have this safety net when accessing the array by index. A runtime error occurs if we try to retrieve a value using an invalid index.

> *If you want to follow along with me, download the repository from GitHub. Open the SwiftCollectionTypes playground from the* **collections-src** *folder. You can find the source code for this demo in the playground page "The Array".*

We can prevent crashes by checking whether the index is within bounds.

```
let index = 1
// check index
if index >= 0,
    index < numbers.count {
    let t = numbers[index]
    print(t)
}
```

If an index is bigger than the size of the array, it's going to point beyond the last item. That's why we need this condition:
```
if index < numbers.count
```

Also, the index must not be a negative number:

```
if index >= 0
```

We can come up with a shorter form. The array has a property called indices that holds the "live" indices of the given array. If it contains the index, then it is valid.

```
// using indices.contains
if numbers.indices.contains(index) {
    let t = numbers[index]
    print(t)
}
```

To push it even further, let's create a custom safe index operator:

```
// Array extension for index bounds check
extension Array {
    subscript (safe index: Index) -> Element? {
        return indices.contains(index) ? self[index] : nil
    }
}
```

We implement the operator in an Array *extension*. Extensions and custom operators are really powerful. They let you add operators to existing types without modifying their code.

Using the newly created operator feels just as natural as using any built-in operator:

```
let emptyArray = [Int]()
let t = emptyArray[safe: index]
print(t ?? "no value at index \(index)")
```

SECTION 6.3
MODIFYING THE ARRAY

So far, we've seen how to access the elements of an array. Now, we'll look at how to modify the array.

> *If you want to follow along with me, download the repository from GitHub. Open the SwiftCollectionTypes playground from the* **collections-src** *folder. You can find the source code for this demo in the playground page "The Array".*

To change the contents of an array after its creation, we must assign it to a variable rather than a constant.

```
var mutableNumbers = [1, 2, 5, 3, 1, 2]
```

We just created a mutable array instance.

The Array exposes different instance methods for adding new values. You can use append() to add a new element to the end of an array:

```
mutableNumbers.append(11)
print(mutableNumbers)
// Output: [1, 2, 5, 3, 1, 2, 11]
```

Here's what happens when we call append(11) on the mutableNumbers array:

append(11)

The new value is appended to the end of the array. If the array was empty, the new element becomes its first element (with index 0).

To insert an element at a given position, use the insert(_:, at:) instance method.

mutableNumbers.insert(42, at: 4)
print(mutableNumbers)
// Output: [1, 2, 5, 3, 42, 1, 2, 11]

The new element is inserted before the element at the current index:
insert(42, **at**: 4)

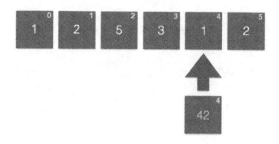

All the elements after the given index are shifted one position to the right.

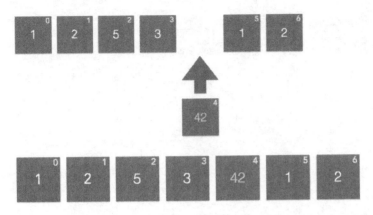

If you pass in the last index, the new element is appended to the array.

insert(77, at: 5)

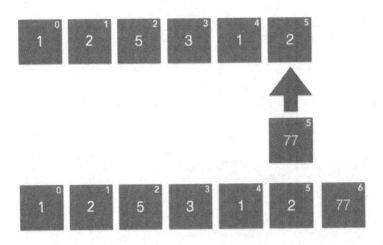

When using insert(_:, at:) make sure that the index is valid. Otherwise, you'll end up in a runtime error.

You can use the remove(at:) instance method to remove an element from an array.

```
mutableNumbers.remove(at: 1)
print(mutableNumbers)
// Output: [1, 5, 3, 42, 1, 2, 11]
```

remove(at: 1)

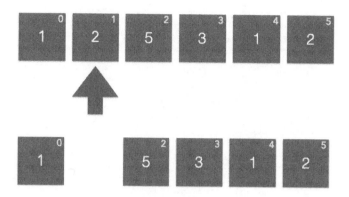

After the element is removed, the gap is closed.

Note that the index we pass to the remove(at:) method must be valid. Otherwise, we cause a runtime error.

Call removeAll() if you want to get rid of all the elements in the array.

```
mutableNumbers.removeAll()
print(mutableNumbers)
// Output: []
```

The method has a keepingCapacity parameter which is set to false by default. You may want to keep the capacity if you plan to reuse the array soon.

```
mutableNumbers = [1, 2, 5, 3, 1, 2]
mutableNumbers.removeAll(keepingCapacity: true)
print("Array count: \(mutableNumbers.count) capacity: \(mutableNumbers.capacity)")
// Output: Array count: 0 capacity: 6
```

removeAll(keepingCapacity: true)

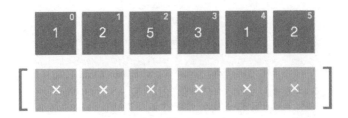

The array won't need to reallocate memory which can be a nice performance optimization.

Although the capacity is kept, we can't reuse the old indices.

Accessing the array by its obsolete indices causes an instant crash!

Rule of thumb
Always check whether the index is out of bounds before accessing it!

The Array has further methods.

- removeFirst() removes and returns the first element,

- whereas removeLast() removes and returns the last element.

```
mutableNumbers = [1, 2, 5, 3, 1, 2]
let wasFirst = mutableNumbers.removeFirst()
print(mutableNumbers)
// Output: [2, 5, 3, 1, 2]

let wasLast = mutableNumbers.removeLast()
print(mutableNumbers)
// Output: [2, 5, 3, 1]
```

Do not call removeFirst() or removeLast() on an empty array, as it will cause - as you may already know - a runtime error.
We talked about some of the most frequently used Array methods. I suggest you download the sample projects and start tinkering with arrays.

Summary

Arrays store values of the same type in an ordered sequence. Choose the array if the order of the elements is important and if the same values shall appear multiple times.

If the order is not important, or the values must be unique, you should rather use a Set.

SECTION 6.4
THE SET

We've seen that the array stores elements in a given order. We can even have duplicates in an array.

What if we need a collection that guarantees the uniqueness of its elements? The Set is the answer.

Sets store unique values with no ordering, and a given value can only appear once. Besides, the Set exposes useful mathematical set operations like union and subtract.

> *If you want to follow along with me, download the repository from GitHub. Open the SwiftCollectionTypes playground from the **collections-src** folder. You can find the source code for this demo in the playground page "The Set".*

We can declare a Set using the following syntax:

```
// Initialize an empty Set
let numbers = Set<Int>()

// Initialize using array literals
let numbers: Set<Int> = [1, 2, 5, 3]
```

Note that we can't use the shorthand form as we did for arrays. If we define it like this:

```
// !!! This declares an Array!
let numbers = [1, 2, 5, 3] // same as let numbers:
Array<Int> = [1, 2, 5, 3]
```

Swift's type inference engine can't figure it out whether we wanted to instantiate a Set or rather an Array. So, it is going to default to Array. We must specify that we need a Set; however, type inference will still work for the values used in the Set. If we initialize the set with an array of Int literals, the type inference engine will infer the type Int:

```
let numbers3: Set = [1, 2, 5, 3]
```

And if we use floating point literals, the inferred type for the values will be Double.

```
let doubles: Set = [1.5, 2.2, 5] // same as -> let doubles:
Set<Double> = [1.5, 2.2, 5]
```

Now, let's clarify the main differences between sets and arrays. By looking at these differences, we'll be able to choose the right one.

The first, fundamental difference is that the Set doesn't allow duplicates. Let's take a look at an example. The following array contains four Int values with the value 1:

```
let onesArray: Array = [1, 1, 1, 1]
print(onesArray)
// Output: [1, 1, 1, 1]
Whereas the Set, declared with the same literals, will only
have one value. The redundant values are skipped.

let onesSet: Set = [1, 1, 1, 1]
print(onesSet)
// Output: [1]
```

The other big difference is that the Set doesn't provide a defined ordering for its elements. For example, the following array will print its contents in the given order:

```
let numbersArray: Array = [1, 2, 3, 4, 5]
print(numbersArray)
// Output: [1, 2, 3, 4, 5]
```

However, for a Set defined with the same values, the order is undefined.

```
let numbersSet: Set = [1, 2, 3, 4, 5]
print(numbersSet)
// Output: undefined order, e.g. [5, 2, 3, 1, 4]
```

We can iterate over the values in a set using a for-in loop.

```
let numbersSet: Set = [1, 2, 3, 4, 5]
// for-in loop
for value in numbersSet {
    print(value)
```

```
}
// Output: undefined order, e.g. 5, 2, 3, 1, 4
```
Remember: the Set doesn't define a specific ordering of its elements. Thus, don't assume anything about the order in which the elements are returned. If you need to iterate in a specific order, call the sorted() method.

```
for value in numbers.sorted() {
    print(value)
}
// Output: 1, 2, 3, 4, 5
```

sorted() returns the elements of the set as an array sorted using the "<" operator.

We can also use the forEach(_:) collection method with sets. This method executes its closure on each element in the Set:

```
numbers.forEach { value in
    print(value)
}
// Output: undefined order, e.g. 5, 2, 3, 1, 4
```

SECTION 6.5
ACCESSING AND MODIFYING THE SET

Now that we know how to create a Set let's talk about accessing and modifying its contents.

> *If you want to follow along with me, download the repository from GitHub. Open the SwiftCollectionTypes playground from the* **collections-src** *folder. You can find the source code for this demo in the playground page "The Set".*

Unlike the array, the set doesn't have indices. We can use the contains() instance method to check whether a value exists in the set:

```
var mutableStringSet: Set = ["One", "Two", "Three"]
let item = "Two"
// set.contains()
if mutableStringSet.contains(item) {
    print("\(item) found in the set")
} else {
    print("\(item) not found in the set")
}
// Output: Two found in the set
```

contains() returns a boolean value, which lets us use it in conditional logic like in this example. If the element cannot be found or if the set is empty, contains() return false.

Regarding empty sets: we can check whether a set has elements through the isEmpty property:

```
let strings = Set<String>()
if strings.isEmpty {
    print("Set is empty")
}
// Output: Set is empty
```

Alternatively, we can use the set's count property. As you might've guessed, count returns the number of elements in the set.

```
let emptyStrings = Set<String>()
if emptyStrings.count == 0 {
    print("Set has no elements")
}
// Output: Set has no elements
```

Note that a set's count and capacity properties can return different values. count returns the number of existing elements in the set. Whereas the capacity shows how many items it can store without allocating new storage. We'll talk about this in a moment.

*Don't confuse **count** with **capacity**!*

Use the insert() Set instance method to add new elements to the set.

```
var mutableStringSet: Set = ["One", "Two", "Three"]
mutableStringSet.insert("Four")
```

To remove elements from a set, we can call the remove() method.

```
mutableStringSet.remove("Three")
```

The call does nothing if the element is not in the list. The remove() method returns the element that was removed from the list. We can use this feature to check whether the value was indeed deleted.

```
mutableStringSet = ["One", "Two", "Three"]
if let removedElement = mutableStringSet.remove("Ten") {
    print("\(removedElement) was removed from the Set")
} else {
    print("\"Ten\" not found in the Set")
}
// Output: "Ten" not found in the Set
```

To remove all the elements in the set call removeAll():

```
mutableStringSet.removeAll()
```

The Set's removeAll() method has a keepingCapacity parameter which is set to false by default. It has the same effect as for arrays.

```
var numbersSet: Set = [1, 2, 5, 3, 1, 2]
numbersSet.removeAll(keepingCapacity: true)
print("count: \(numbersSet.count) capacity:
```

```
\(numbersSet.capacity)")
// count: 0 capacity: 6
```

By setting the value of keepingCapacity to true, we tell the compiler to keep the set's memory after removing its elements. The array won't need to reallocate memory upon its next usage.

SECTION 6.6
SET OPERATIONS

The Set exposes useful methods that let us perform fundamental operations.

> *If you want to follow along with me, download the repository from GitHub. Open the SwiftCollectionTypes playground from the **collections-src** folder. You can find the source code for this demo in the playground page "The Set".*

Union

union() creates a new set with all the elements in the two sets. If the two sets have elements in common, only one instance will appear in the resulting set.

```
let primes: Set = [3, 5, 7, 11]
let odds: Set = [1, 3, 5, 7]

// set.union(otherSet)
let union = primes.union(odds)
print(union.sorted())
// Output: [1, 3, 5, 7, 11]
```

Note that 3, 5 and 7 can be found in both sets, but the resulting set will not contain duplicates.

Intersection

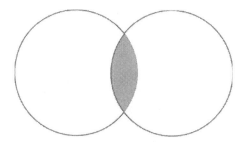

The result of calling the intersection() Set instance method is a set which holds the elements that appear in both sets.

```
let primes: Set = [3, 5, 7, 11]
let odds: Set = [1, 3, 5, 7] let intersection =
primes.intersection(odds)
print(intersection.sorted())
// Output: [3, 5, 7]
```

Subtract

We can also subtract one set from another.

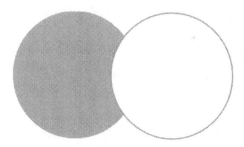

The result will contain those values which are only in the source set and not in the subtracted set.

```
let primes: Set = [3, 5, 7, 11]
let odds: Set = [1, 3, 5, 7] let subtract =
primes.subtracting(odds)
print(subtract.sorted())
// Output: [11]
```

Symmetric Difference

The symmetricDifference() method returns a Set with the elements that are only in either set, but not both.

The Set exposes many other useful methods, like the ones which let us test for equality and membership. I suggest you download the sample projects and start experimenting with sets.

SECTION 6.7
THE HASHABLE PROTOCOL

Why do we need to talk about the Hashable protocol? You'll see in a minute.

> *If you want to follow along with me, download the repository from GitHub. Open the SwiftCollectionTypes playground from the* **collections-src** *folder. You can find the source code for this demo in the playground page "The Hashable Protocol".*

First, I'm going to create a very simple structure.

```
struct SimpleStruct {
    var identifier: String
}
```

With our struct in place, we could create for example an array:

```
var array = [SimpleStruct]()
array.append(SimpleStruct(identifier: "id"))
```

This code compiles just fine.
Let's try to do the same but this time I'm going to declare a Set.

```
var set = Set<SimpleStruct>()
```

If we try to compile this code we get an error. The compiler tells that our type doesn't conform to the Hashable protocol.

What does that mean? The Set is a special kind of a Swift collection type. Unlike arrays, a set cannot have duplicates. In order to check for duplicates the Set must be of a type that has a way to tell whether an instance is unique. Swift uses a hash value for this purpose.

The hash value is a unique value of type Int which must be equal if two values are the same. In Swift, all basic built-in types are hashable, so we can use a String, a Bool, an Int or a Double in a set.

So, if we declare the Set like this:

```
var set = Set<Double>()
```

The compiler won't complain anymore, since Double provides a hash value.

To make our SimpleStruct work with the Set, we must make it conform to the Hashable protocol. First. we add the Hashable conformance:

```
struct SimpleStruct: Hashable {
    var identifier: String
}
```

Now, let's take a look at the Hashable protocol.

```
public protocol Hashable : Equatable {

    /// The hash value.
    ///
    /// Hash values are not guaranteed to be equal across different executions of
    /// your program. Do not save hash values to use during a future execution.
    public var hashValue: Int { get }
}
```

It has a single read-only, gettable property called hashValue. This hash value must be unique. SimpleStruct has only one property of type String. String is a built-in type which already implements the Hashable protocol. Thus, implementing the hashValue property becomes a straightforward task.

```
struct SimpleStruct: Hashable {
    var identifier: String

    public var hashValue: Int {
        return identifier.hashValue
    }
}
```

If you try to compile the code, you'll still have a problem: SimpleStruct must also confirm to the Equatable protocol.
Hashable inherits from the Equatable protocol. If a protocol inherits from another one, all conforming types must also implement the requirements defined in that protocol.

Conforming to the Equatable protocol is straightforward, too. We have to implement the "==" operator. The equality operator is a static method that tells whether two instances of the given type are equal or not. We consider two SimpleStruct instances to be equal if their identifiers are equal.

```
struct SimpleStruct: Hashable {
    var identifier: String

    public var hashValue: Int {
        return identifier.hashValue
    }

    public static func == (lhs: SimpleStruct, rhs: SimpleStruct) -> Bool {
        return lhs.identifier == rhs.identifier
    }
}
```

Now, our code will compile without errors.

We must adopt the Hashable protocol to ensure that the given value is unique. This is required if we want to use our type in a set or as keys for a dictionary.

SECTION 6.8
THE DICTIONARY

The Dictionary, also known as hash-map, stores key-value pairs. Use this collection type if you need to look up values based on their identifiers.

Each value must be associated with a key that is unique. The order of the keys and values is undefined.

Just like the other Swift collection types, the Dictionary is also implemented as a generic type.

SECTION 6.9
CREATING DICTIONARIES

If you want to follow along with me, download the repository from GitHub. Open the SwiftCollectionTypes playground from the **collections-src** *folder. You can find the source code for this demo in the playground page "The Dictionary".*

To create a dictionary, we must specify the key and the value type. We can create an empty dictionary like this:

```
// Specify the key and the value type to create an empty
dictionary
var dayOfWeek = Dictionary<Int, String>()
```

Another way is to use the initializer syntax:

```
var dayOfWeek2 = [Int: String]()
```

We can initialize a dictionary with a dictionary literal:

```
var dayOfWeek3: [Int: String] = [0: "Sun", 1: "Mon", 2: "Tue"]
```

Swift can infer the type of the keys and the values based on the literal:

```
var dayOfWeek4 = [0: "Sun", 1: "Mon", 2: "Tue"]
```

SECTION 6.10
HETEROGENEOUS DICTIONARIES

When creating a dictionary, the type of the keys and values is supposed to be consistent - e.g. all keys are of Integer type and all the values are of type String.

Type inference won't work if the type of the dictionary literals is mixed.

```
var mixedDict = [0: "Zero", 1: 1.0, "pi": 3.14]

// Error: Heterogeneous collection literal could only be
inferred to '[AnyHashable : Any]'; add explicit
```

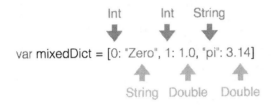

The compiler asks us to explicitly add type annotations if we really want to declare a heterogeneous dictionary. And, in certain cases, you may want to do that. For instance, when converting JSON payloads or property lists, a typed dictionary won't work. To create a heterogeneous dictionary, the keys must conform to the Hashable protocol.

As we've seen previously, the Hashable protocol defines a property requirement called hashValue. The hashValue property must return a unique value for a given instance. No two different instances shall have the same hash value.

The Swift standard library has a type called AnyHashable. AnyHashable can hold a value of any type conforming to the Hashable protocol. This type can be used as the super-type for keys in heterogeneous dictionaries.

So, we can create a heterogeneous collection like this:

```
var mixedDict: [AnyHashable: Any] = [0: "Zero", 1: 1.0,
"pi": 3.14]
```

AnyHashable is a structure which lets us create a type-erased hashable value that wraps the given instance. We could also define the mixedDict dictionary by wrapping the key values in AnyHashable instances and by explicitly declaring the values as type Any:

```
var mixedDict = [AnyHashable(0): "Zero" as Any,
                 AnyHashable(1): 1.0 as Any,
                 AnyHashable("pi"): 3.14 as Any]
```

Although this version compiles just fine, the shorthand syntax is obviously shorter and more human friendly.
AnyHashable has a base property that represents the wrapped value. It can be cast back to the original type using the as, conditional as? or forced as! cast operators.

```
let piWrapped = AnyHashable("pi")
if let unwrappedPi = piWrapped.base as? String {
    print(unwrappedPi)
}
```

SECTION 6.11
ACCESSING & MODIFYING
THE CONTENTS OF A DICTIONARY

We can access the value associated with a given key using the subscript syntax:

```
var dayOfWeek = [0: "Sun", 1: "Mon", 2: "Tue"]
if let day = dayOfWeek[2] {
    print(day)
}
// Prints: Tue
```

We can also iterate over the key-value pairs of a dictionary using a for-in loop. Since it's a dictionary, items are returned as a key-value tuple:

```
for (key, value) in dayOfWeek {
    print("\(key): \(value)")
}
```

The dictionary's keys property provides access to its keys:

```
for key in dayOfWeek.keys {
    print(key)
}
```

And the values property will return its values:

```
for value in dayOfWeek.values {
    print(value)
}
```

To add a new item to the dictionary, use a new key as the subscript index, and assign it a new value:

```
dayOfWeek[3] = "Wed"
print(dayOfWeek)
// Prints: [2: "Tue", 0: "Sun", 1: "Mon", 3: "Wed"]
```

We can update an existing item using the subscript syntax with an existing key:

```
dayOfWeek[2] = "Mardi"
print(dayOfWeek)
// Prints: [2: "Mardi", 0: "Sun", 1: "Mon", 3: "Wed"]
```

Alternatively, we can call the updateValue(_: forKey:) method.

```
dayOfWeek.updateValue("Tue", forKey: 2)
```
If the key doesn't exist, it adds a new key-value pair to the dictionary. You can remove a value from a dictionary by assigning nil for its key:

```
dayOfWeek[1] = nil
```

You can achieve the same result (with more typing) by calling the removeValue(forKey:) method:

```
dayOfWeek.removeValue(forKey: 2)
```

Invoke removeAll() if you need to wipe out the dictionary:

```
dayOfWeek.removeAll()

print(dayOfWeek)
// Output: [:]
```

removeAll() has a keepingCapacity parameter, which is set to false by default. If you pass true, the operation preserves the capacity of the underlying buffer.

```
dayOfWeek = [0: "Zero", 1: "One", 10: "Ten"]
dayOfWeek.removeAll(keepingCapacity: true)

print("\(dayOfWeek), count: \(dayOfWeek.count) capacity: \(dayOfWeek.capacity)")
// Output: [:], count: 0 capacity: 3
```

This is a performance optimization if you plan to reuse the dictionary.

CHAPTER 7
BASIC SORTING

Understanding the inner workings and knowing how to implement the basic sorting algorithms gives you a strong foundation to building other, more sophisticated algorithms.

We're going to analyze how each algorithm works and we'll implement them from scratch using Swift.

> *What is sorting first of all? Sorting is a technique for arranging data in a logical sequence according to some well-defined rules.*

Working with sorted data is way more efficient than accessing elements in an unordered sequence. Sorting plays a key role in commercial data processing and scientific computing.

We're going to start our journey into the area of sorting algorithms by studying three elementary sorting methods:

- Selection Sort
  Works by selecting the smallest item and replacing it with the previous one. This algorithm has a quadratic - $O(n^2)$ - time complexity, and it is one of the simplest sorting algorithms.

- Insertion Sort

    As its name states, works by inserting elements into their proper place; to make space for the current item, larger elements must move one position to the right.

    Insertion sort has quadratic - $O(n^2)$ - worst time complexity; however, the performance of the insertion sort is largely affected by the initial order of the elements in the sequence.

- Bubble Sort

    This algorithm works by repeatedly evaluating adjacent items and swapping their position if they are in the wrong order. The Bubble sort is easy to implement, but it's also quite inefficient: it's average and worst-case complexity are both quadratic - $O(n^2)$.

SECTION 7.1
SELECTION SORT

Selection Sort is one of the simplest sorting algorithms.

It starts by finding the smallest item and exchanging it with the first one. Then, it finds the next smallest item and exchanges it with the second item. The process goes on until the entire sequence is sorted.

Implementation

> *If you want to follow along with me, download the repository from GitHub and open the SelectionSort playground from the **basic-sorting-src** folder.*

Here's the selection sort algorithm implemented in Swift.

```swift
func selectionSort(_ input: [Int]) -> [Int] {
    guard input.count > 1 else {
        return input
    }
    
    var result = input
    
    for index in 0..<(result.count - 1) {
        var indexLowest = index
        
        for forwardIndex in (index + 1)..<result.count {
            if result[forwardIndex] < result[indexLowest] {
                indexLowest = forwardIndex
            }
        }
        
        if index != indexLowest {
            result.swapAt(index, indexLowest)
        }
    }
    return result
}
```

The selectionSort() function takes an array of integers as input and returns the sorted copy of the original array.

```swift
func selectionSort(_ input: [Int]) -> [Int] {...}
```

Sorting makes sense only if the array has at least two elements; otherwise, we just return a copy of the original array.

```
func selectionSort(_ input: [Int]) -> [Int] {
    guard input.count > 1 else {
        return input
    }

    var result = input

    // ...
    return result
}
```

We are going to sort the contents of this copy and return it from our function.

```
func selectionSort(_ input: [Int]) -> [Int] {
    guard input.count > 1 else {
        return input
    }

    var result = input

    for index in 0..<(result.count - 1) {
        var indexLowest = index

        for forwardIndex in (index + 1)..<result.count {
            if result[forwardIndex] < result[indexLowest] {
                indexLowest = forwardIndex
            }
        }

        if index != indexLowest {
            result.swapAt(index, indexLowest)
        }
    }
    return result
}
```

The function has two loops. The outer loop iterates through each element, and it maintains the index which denotes the sorted portion of the array. The inner loop finds the lowest number in the rest of the array. For each element, the function swaps positions with the lowest value from the remainder of the array.

Given the unsorted array [1, 2, 4, 3, 0] the outer loop picks the first element, which is 1.[1, 2, 4, 3, 0]

Then, the inner loop finds the lowest number in the rest of the array. Since 0 is smaller than 1, the position of the two elements gets swapped.

Now, we have the first element in the sorted portion of the array.[1, 2, 4, 3, 0]

[0, 2, 4, 3, 1]

The outer loop picks the next element: 2

[0, 2, 4, 3, 1]

The inner loop finds the lowest number starting from the next position. 1 is smaller than 2, so their position gets swapped. Our array has now two sorted elements:

[0, 1, 4, 3, 2]

4 is picked by the outer loop next.

[0, 1, 4, 3, 2]

The inner loop finds the lowest number, which is 2. Since 2 is smaller than 4, they are swapped, and we get:

[0, 1, 2, 3, 4]

The outer loop picks the next element, which is 3.

[0, 1, 2, 3, 4]

The rest of the array consists of only one element. 4 is not smaller than 3, so their position is not swapped.

Finally, we get our sorted array

[0, 1, 2, 3, 4]

Selection Sort Time Complexity

The algorithm sorts the array from left to right. The running time is affected by the number of swaps and compares. The number of exchanges depends on the original order of the elements.

The worst case is when the array is reverse-sorted: the number of swaps will be equal to the number of elements in the array. However, the time complexity of the selection sort is dominated by the number of compares.

Let's inspect the source code to find the number of compares:

```
for index in 0..<(result.count - 1) {
    var indexLowest = index

    for forwardIndex in (index + 1)..<result.count {
        if result[forwardIndex] < result[indexLowest] {
            indexLowest = forwardIndex
        }
    }

    if index != indexLowest {
        result.swapAt(index, indexLowest)
    }
}
```

For an array with N elements, the outer loop iterates n - 1 times (from index 0 to index (n - 2)):

```
for index in 0..<(result.count - 1) {...}
```

The inner loop iterates (n - 2) times first, then (n - 3) times, and so on. When the outer loop reaches the penultimate index, the inner loop only iterates once.

This means that we have (n - 1) + (n - 2) + ... + 1 iterations in total.

If we add the number of *worst-case swaps*, the time complexity becomes

Swaps	Loops
n	(n - 1) + (n - 2) + ... + 1

Which gives:

Swaps + Loops

$$1 + 2 + \ldots (n-2) + (n-1) + n = \frac{n \times (n+1)}{2}$$

So, the selection sort algorithm's worst time complexity is:

$$\frac{n^2 + n}{2}$$

If you are wondering how did we get to this formula, I'd suggest you revisit Calculate Sum(n).

If the data is already sorted, the number of swaps will be 0.

Swaps	Loops
0	$(n-1) + (n-2) + \ldots + 1$

~~Swaps~~ + Loops

$$1 + 2 + \ldots (n-2) + (n-1) = \frac{(n-1) \times n}{2}$$

In this case, the selection sort has the best performance. Its time complexity becomes:

$$\frac{n^2 - n}{2}$$

The demo includes performance measurements. We sort three arrays with 10, 100 and 1000 elements, respectively. The results displayed in the console confirm the quadratic time complexity of the selection sort algorithm.

Next, we inspect the performance of the selection sort with already sorted input. As you'll see, the performance is slightly better with sorted input. To reproduce the worst case, we use arrays sorted in reverse order as input. These will produce the worst running times.

To summarize:

Already Sorted Input	Random Input	Reverse Ordered Input
$\dfrac{n^2 - n}{2}$	$\dfrac{n^2 + (n - i)}{2}$	$\dfrac{n^2 + n}{2}$
	$0 < i < n$	

A big disadvantage of the selection sort is that it's insensitive to input. It takes almost as long to run selection sort on an array that is already in order as it does for a randomly ordered array.

The selection sort is definitely easy to grasp and implement. As it relies on nested loops, its time complexity is quadratic.

The running time of the selection sort is mostly affected by the number of compares performed by the algorithm. The running time of the selection sort won't vary too much whether you sort an already sorted sequence or a totally shuffled one.

As a conclusion: understanding the selection sort and knowing how to implement it is important; however, try to avoid it in real code.

SECTION 7.2
INSERTION SORT

Insertion sort is a basic sorting algorithm, which works by analyzing each element and inserting it into its proper place, while larger elements move one position to the right.

Insertion sort has quadratic time complexity. However, the performance of the insertion sort is largely affected by the initial order of the elements in the sequence.

Implementation
In the following demo, we are going to implement the insertion sort algorithm in Swift. We'll visualize how insertion sort works. Then, we are going to analyze the time complexity of this algorithm.

We will conduct an interesting experiment: we'll compare the efficiency of the insertion sort with the selection sort algorithm that was presented in the previous episode. There will be three distinct use-cases: first, we'll use a shuffled array as input, then a partially sorted one, and finally an already sorted array.

> *If you want to follow along with me, download the repository from GitHub and open the InsertionSort playground from the* **basic-sorting-src** *folder.*

The insertionSort() function takes an array of integers as input and returns the sorted copy of the original array.

```
func insertionSort(_ input: [Int]) -> [Int] {
    var result = input

    let count = result.count

    for sortedIndex in 1 ..< count {
        var backIndex = sortedIndex
        while backIndex > 0 && result[backIndex] < result[backIndex - 1] {
            result.swapAt((backIndex - 1), backIndex)
```

```
            backIndex -= 1
        }
    }
    return result
}
```

We clone the input array first. This copy will hold the sorted result. The insertionSort() function uses two loops. The outer loop progresses as we process the array.

```
    let count = result.count

    for sortedIndex in 1 ..< count {
        var backIndex = sortedIndex
        while backIndex > 0 && result[backIndex] < result[backIndex - 1] {
            result.swapAt((backIndex - 1), backIndex)
            backIndex -= 1
        }
    }
```

We start at index 1 because the inner loop has to decrement the index as it traverses the subarray backward.

Note that we don't need to validate the input size as we did in the implementation of the selectionSort function. The outer loop counter starts at one and ends at inputSize, therefore empty or single element arrays won't be processed anyway.

For a sorted array, the number at the current index must be bigger than all previous ones. The inner loop steps backward through the sorted sub-array and swaps the values if it finds a previous number that is larger than the one at the current index.

Given the unsorted array[1, 2, 4, 3, 0] the outer loop picks the element at index 1, which is 2.[1, 2, 4, 3, 0]

The outer loop starts at index 1 because the inner loop must traverse the array backward. The outer index also marks the sorted portion of the array, which, at this point, consists of only one element:

Sorted [1] Unsorted [2, 4, 3, 0]

The inner loop compares the number at the current index with the elements in the sorted portion of the array. That is, it compares 2 with 1:

1 | 2, 4, 3, 0

Since 2 is bigger than 1, the numbers are not swapped.

At this point, the sorted portion includes the numbers [1, 2] and the unsorted has the numbers [4, 3, 0].

The outer loop picks the element at index 2, which is 4.

1, 2 | 4, 3, 0

The inner loop compares 4 with the elements in the sorted part: 4 is not smaller than 2. Therefore, no swap occurs.

The sorted and unsorted portions are now as follows:

Sorted [1, 2, 4] Unsorted [3, 0]

The outer index gets incremented; the next element is the number 3.

1, 2, 4 | 3, 0

The inner loop performs the comparisons with the sorted portion. 3 is smaller than 4, so they are swapped. 3 is not less than 2, so there are no further swaps. We have now [1, 2, 3, 4] in the sorted part.

Sorted [1, 2, 3, 4] Unsorted [0]

The outer index picks the last element from the unsorted portion of the array.

1, 2, 3, 4 | 0

Since 0 is smaller than all the numbers in the sorted portion, it gets repeatedly swapped with each item from the sorted sub-array.

1, 2, 3, 0, 4

1, 2, 0, 3, 4

1, 0, 2, 3, 4

0, 1, 2, 3, 4

The unsorted part is empty, so now we have all the elements in order:

Sorted [0, 1, 2, 3, 4] Unsorted []

Insertion Sort Time Complexity
Insertion sort uses two nested loops. Therefore, its worse time complexity is quadratic. However, it is much more performant than the selection sort if the input array includes already sorted sequences.

Unlike the selection sort, which is insensitive to input, the running times of the insertion sort algorithm can vary considerably depending on the order of the input array.

The best case is when the array is already sorted. During each iteration, the next element from the unsorted portion is only compared with the rightmost element of the sorted subsection of the array.

0, 1, 2, 3, 4

0 | 1, 2, 3, 4

0, 1 | 2, 3, 4

0, 1, 2 | 3, 4

0, 1, 2, 3 | 4

0, 1, 2, 3, 4

No swaps are made since the elements are already in place.

For an already ordered array which has N elements, the algorithm will execute N - 1 compares and 0 exchanges.

Swaps   Compares
  0   n - 1

In other words, the best running time of the insertion sort is linear - O(n).

The worst case is when the elements are in reverse order.

4, 3, 2, 1, 0

In this case, every iteration of the inner loop will compare the current element with all the elements of the sorted part.

						Swaps
4 | 3, 2, 1, 0						
4, 3 | 2, 1, 0	3, 4 | 2, 1, 0					1
3, 4, 2 | 1, 0	3, 2, 4 | 1, 0	2, 3, 4 | 1, 0				2
2, 3, 4, 1 | 0	2, 3, 1, 4 | 0	2, 1, 3, 4 | 0	1, 2, 3, 4 | 0			3
1, 2, 3, 4, 0	1, 2, 3, 0, 4	1, 2, 0, 3, 4	1, 0, 2, 3, 4	0, 1, 2, 3, 4		4

The number of swaps will be equal to the number of items in the sorted subsection. To calculate the time complexity, we need to sum up the number of compares and the number of swaps:

Swaps   Compares
1 + 2 + ... + (n-1)   1 + 2 + ... + (n-1)

$$\frac{(n-1) \times n}{2} + \frac{(n-1) \times n}{2} = n^2 - n$$

So, the worst case complexity of the insertion sort is $n^2 - n$.

When using Big-O notation, we discard the low-order term which gives $O(n^2)$ - quadratic running time.

The average case when each element is halfway in order, the number of swaps and compares is halved compared to the worst case. This gives us $\dfrac{n^2 - n}{2}$, which is also a quadratic time complexity.

To summarize: the insertion sort performs in linear time for already or almost sorted arrays. When the input is shuffled or in reverse order, the insertion sort will run in quadratic time.

Already Sorted Input	Random Input	Reverse Ordered Input
n - 1	$\dfrac{n^2 - n}{2}$	n² - n
O(n)	O(n²)	O(n²)

Go and check out the performance measurements I included in the demo. The tests are executed on shuffled arrays and already sorted ones. The results will confirm that insertion sort performs in linear time when the input is sorted. For random input, the running time of the insertion sort is $O(n^2)$.

Insertion Sort vs. Selection Sort

Let's conduct an interesting experiment. Let's compare the two sorting algorithms we've been studying so far.

> *If you want to follow along with me, download the repository from GitHub. Open the InsertionSort playground from the* **basic-sorting-src** *folder and scroll down to "Selection Sort vs. Insertion Sort".*

First, we'll use shuffled arrays.

```
inputSize = 1000
let random1000 = Array<Int>.randomArray(size: inputSize,
maxValue: 1000)
execTime = BenchTimer.measureBlock {
    _ = insertionSort(random1000)
}
print("\nAverage insertionSort() execution time for
\(inputSize) elements: \(execTime.formattedTime)")

execTime = BenchTimer.measureBlock {
    _ = selectionSort(random1000)
}
print("Average selectionSort() execution time for
\(inputSize) elements: \(execTime.formattedTime)")
```

As the input size increases, the insertion sort performs slower. The two algorithms are supposed to run in quadratic time, but the selection sort runs 3 times quicker for 100 random numbers and about 20 times faster for 1000 elements.

The insertion sort usually makes fewer comparisons than the selection sort; however, the selection sort requires fewer swaps. As we've seen in the section about the selection sort, the maximum number of swaps is equal to the input size - in other words, it grows linearly with the input in the worst case.

Insertion sort will usually perform $O(n^2)$ swaps. Since writing to memory is usually significantly slower than reading, the selection sort may perform better than the insertion sort for larger input.

INTRODUCTION TO ALGORITHMS AND DATA STRUCTURES IN SWIFT 4

The situation changes drastically if we run our tests with sorted input.

```
inputSize = 1000
let progressiveArray1000 =
Array<Int>.incrementalArray(size: inputSize)
execTime = BenchTimer.measureBlock {
    _ = insertionSort(progressiveArray1000)
}
print("\nAverage insertionSort() execution time for
\(inputSize) elements: \(execTime.formattedTime)")

execTime = BenchTimer.measureBlock {
    _ = selectionSort(progressiveArray1000)
}
print("Average selectionSort() execution time for
\(inputSize) elements: \(execTime.formattedTime)")
```

The best-case, linear nature of the insertion sort shows its benefits over the selection sort algorithm, which is insensitive to input.
Because the array is already sorted, the insertion sort runs its best-case scenario. There won't be any swaps, and the number of compares is n - 1. The insertion sort algorithm runs in linear time, whereas the selection sort algorithm runs in quadratic time. In our example, this translates to almost 10 times slower performance compared to the insertion sort.

To sum up the running times for the insertion sort:

- Best case is linear $O(n)$ - for almost sorted input

- Average case is quadratic $O(n^2)$ - for shuffled input

- Worst case is also quadratic $O(n^2)$ - for reverse sorted input

The insertion sort algorithm is easy to understand and implement, and its worst-case time complexity is quadratic. However, its running time decreases if the input contains already sorted elements. In the best case, when the input is already ordered, the insertion sort performs only n - 1 compares, where n is the input size.

Unlike the selection sort's running time, which - as we have seen - is insensitive to input, the insertion sort performs better when the input is partially or totally ordered.

As a conclusion: in spite of its simplicity, the insertion sort may perform surprisingly well with partially sorted sequences. In fact, even Apple relies on insertion sort in their sort() implementation if the range to be sorted contains fewer than 20 elements.

Swift is open-source, so you can check the implementation of the sort function in the Swift Github repository:
https://github.com/apple/swift/blob/master/stdlib/public/core/Sort.swift.gyb

SECTION 7.3
BUBBLE SORT

The Bubble Sort algorithm works by repeatedly evaluating adjacent items and swapping their position if they are in the wrong order.

In the following demo, we are going to implement the bubble sort algorithm. As with the other algorithms, we are going to analyze the time complexity of the Bubble sort, and visualize how it works. Then, we are going to compare the Bubble sort, the insertion sort, and the selection sort algorithm in terms of efficiency.

Implementation

The bubbleSort() function takes an array of integers as input and returns the sorted copy of the input array.

> *If you want to follow along with me, download the repository from GitHub and open the BubbleSort playground from the **basic-sorting-src** folder.*

```
func bubbleSort(_ input:[Int]) -> [Int] {
    guard input.count > 1 else {
        return input
    }
    
    var result = input
    let count = result.count
    
    var isSwapped = false
    
    repeat {
        isSwapped = false
        for index in 1..<count {
            if result[index] < result[index - 1] {
                result.swapAt((index - 1), index)
                isSwapped = true
            }
        }
    } while isSwapped
    
    return result
}
```

The function repeatedly iterates through the array and compares every adjacent pair; if the numbers are not in the right order, their position gets swapped. The process continues in passes until the entire sequence is sorted. The lower values bubble to the beginning of the array.

The isSwapped variable is used to track whether any swaps were made during a pass. If no swap occurred during a pass, it means that the values are in order, and we exit the outer do-while loop.

Let's see the bubble sort in action.

We start with the unsorted array

[1, 4, 2, 3, 0]

This is the first pass:

Bubble Sort Pass #1

1, 4, 2, 3, 0 ➡ **1, 4**, 2, 3, 0

1, **4, 2**, 3, 0 ➡ **1, 2, 4**, 3, 0

1, 2, **4, 3**, 0 ➡ 1, **2, 3, 4**, 0

1, 2, 3, **4, 0** ➡ **1, 2, 3, 0, 4**

The first two elements are checked. Since 1 is smaller than 4, their order is kept. The next pair is 4 and 2. Because 4 is bigger than 2, their positions get swapped. Then, we compare 4 and 3. They are not in order, so they are swapped. The last pair in this pass is 4 and 0. Since 0 is smaller than 4, their position is exchanged.

We continue with the second pass.

Bubble Sort Pass #2

1, 2, 3, 0, 4 ➡ **1, 2**, 3, 0, 4

1, **2, 3**, 0, 4 ➡ **1, 2, 3**, 0, 4

1, 2, **3, 0**, 4 ➡ 1, **2, 0, 3**, 4

1, 2, 0, **3, 4** ➡ **1, 2, 0, 3, 4**

The first three elements are already in order. Therefore, the algorithm does not swap them. The first pair which requires swapping is found at index 2 and 3. The last pair is already in order, so we keep them at their current position.

The third pass bubbles the value 0 until it almost reaches the beginning of the list.

Bubble Sort Pass #3

1, 2, 0, 3, 4 ➡ **1, 2**, 0, 3, 4
1, **2, 0**, 3, 4 ➡ 1, **0, 2**, 3, 4
1, 0, **2, 3**, 4 ➡ 1, 0, **2, 3**, 4
1, 0, 2, **3, 4** ➡ 1, 0, 2, **3, 4**

However, the array is not yet sorted, so we need another pass.

Bubble Sort Pass #4

1, 0, 2, 3, 4 ➡ **0, 1**, 2, 3, 4
0, **1, 2**, 3, 4 ➡ 0, **1, 2**, 3, 4
0, 1, **2, 3**, 4 ➡ 0, 1, **2, 3**, 4
0, 1, 2, **3, 4** ➡ 0, 1, 2, **3, 4**

The fourth pass only exchanges the very first pair of numbers: now 0 is finally at the beginning of the sequence. The rest of the array is already sorted, *but the algorithm is not aware of this*.
So the algorithm executes another pass.

Bubble Sort Pass #5

0, 1, 2, 3, 4 ➡ **0, 1**, 2, 3, 4
0, **1, 2**, 3, 4 ➡ 0, **1, 2**, 3, 4
0, 1, **2, 3**, 4 ➡ 0, 1, **2, 3**, 4
0, 1, 2, **3, 4** ➡ 0, 1, 2, **3, 4**

The fifth and final pass does not find any pairs that need to be swapped.

Bubble Sort Time Complexity

Now that we understand the inner workings of the bubble sort algorithm let's analyze its time complexity.

If the input is already sorted, the bubble sort needs only one pass, which executes n - 1 compares. All the elements stay in place, so the number of swaps will be 0.

Swaps	Compares
0	n - 1

This means that the running time of the bubble sort only depends on the input size when the sequence is already sorted. In other words, the best-time complexity of the bubble-sort is linear.

The worst case is when the array is reverse-sorted. If there are n items in the sequence, the algorithm will run n passes in total. During each pass, our function executes n-1 comparisons. This means $n \times (n - 1)$ compares.

For a reverse-ordered sequence, the number of swaps will be n - 1 in the first pass, n - 2 in the second pass, and so on, until the last exchange is made in the penultimate pass.

Swaps	Compares
(n-1) + (n-2) +... + 2 + 1	n x (n-1)

The total number of swaps is $\dfrac{(n - 1) \times n}{2}$. The worst-case running time of our bubble sort implementation is the sum of the swaps and compares.

Swaps	Compares
$\dfrac{(n-1) \times n}{2}$	(n-1) x n

This is clearly an order of n squared time complexity.

$$\text{Swaps + Compares} = \frac{3 \times (n^2 - n)}{2}$$

The average time complexity of the bubble sort is also quadratic.

To summarize, the bubble sort has a linear running time for already sorted arrays and runs in order of n squared average and worst time complexity.

These running times might look similar to the time complexities of the insertion sort algorithm. However, the bubble sort needs considerably more swaps than the insertion sort. The higher number of swaps will result in slower performance.

Therefore, the insertion sort performs considerably better than the bubble sort.

Due to its poor performance, the bubble sort is almost never used in real software; however, because it's easy to grasp and implement, the bubble sort algorithm is often used in introductory computer science materials.

Bubble vs. Insertion vs. Selection Sort

Now, let's execute some performance measurements.

```
inputSize = 100
let random100 = Array<Int>.randomArray(size: inputSize,
maxValue: 100)
execTime = BenchTimer.measureBlock {
    _ = insertionSort(random100)
}
print("\nAverage insertionSort() execution time for
\(inputSize) elements: \(execTime.formattedTime)")

execTime = BenchTimer.measureBlock {
    _ = selectionSort(random100)
}
print("Average selectionSort() execution time for
\(inputSize) elements: \(execTime.formattedTime)")

execTime = BenchTimer.measureBlock {
    _ = bubbleSort(random100)
}
print("Average bubble() execution time for \(inputSize)
elements: \(execTime.formattedTime)")
```

For 10 items there is no noticeable difference between the three algorithms. The bubble sort algorithm starts to show its weakness when sorting 100 random numbers. Its performance gets worse as the input size increases.

The Bubble sort is easy to implement, but it has the worst performance among the basic sorting algorithms. The high number of swaps are to blame for the low performance of the bubble sort.
Although we could slightly optimize our existing bubble sort implementation, the insertion sort is going to outperform even the optimized version for larger input.

> *You should avoid bubble sort in production code. Use insertion sort if you need a simple yet acceptable solution.*

In this chapter we analyzed some of the basic sorting algorithms. Studying them is definitely worth the effort to deepen our knowledge in algorithms.

Next, we'll analyze sorting algorithms that can run way faster than any of the elementary algorithms we've studied so far.

CHAPTER 8
ADVANCED SORTING

In this chapter, we're going to take a look at two advanced sorting algorithms.

The merge sort and quick sort are efficient and can be used in production code. These sorting algorithms are actually included in various libraries and frameworks.

The merge sort splits the sequence to be sorted into two halves. Then, it sorts the halves. The sorted parts get combined. During this merge step, additional sorting is done. Finally, we get the sorted result.

The other sorting algorithm we'll be studying is the quicksort. This algorithm uses a similar approach like the merge sort - also known as *divide-and-conquer* technique.

The difference is that the resulting halves are already sorted before the merge. So, there's no need for further sorting when combining the parts during the last step.

All right, now let's delve into these algorithms.

SECTION 8.1
THE MERGE SORT

Let's visualize the merge sort to make it easier to understand how it works. This is the array we want to sort:

[0, 9, 6, 2, 3, 2, 1, 3]

First, we split the array into two parts.

[0, 9, 6, 2, 3, 2, 1, 3]

[0, 9, 6, 2]

The array has 8 elements, so the split index is 4. We got two sublists, each with 4 elements. We continue splitting the halves until we get to the single-element sublists.

First, we split the left half. After splitting it, we get two sublists, each with two elements.

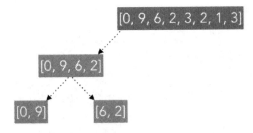

Next, we split these, too.

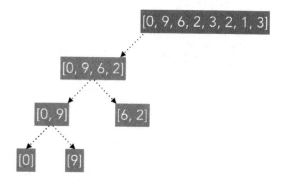

We cannot split the left side further, since now we have only one-element arrays.

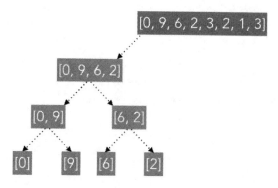

The one-element arrays are sorted and merged.

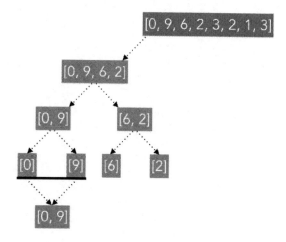

As a result, we'll have 2-element sublists which are ordered.

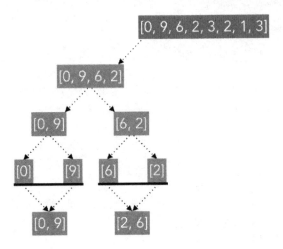

One more merge and sort, and we have the elements of the left half in the desired order.

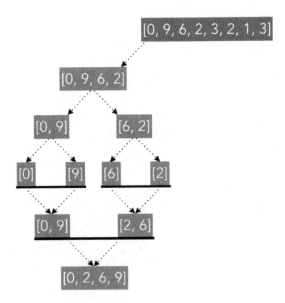

So, now comes the sorting and merging phase. After two steps, the elements of the left half are ordered.

We then follow the same steps for the right half:

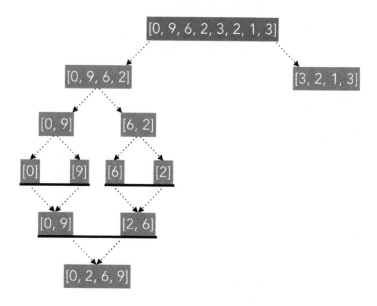

One split, then two more splits on the 2-element sublists, and we get the one-element arrays.

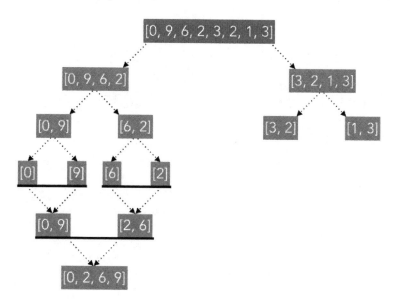

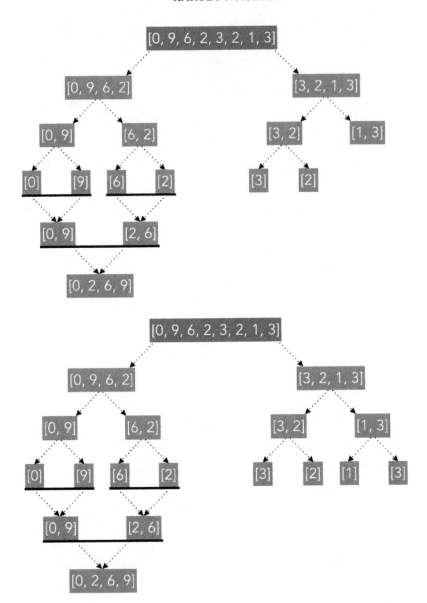

Next, the single-element sublists are sorted and combined until the right half is also sorted.

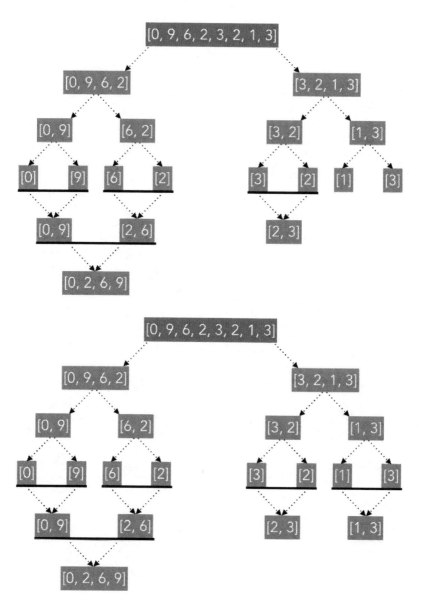

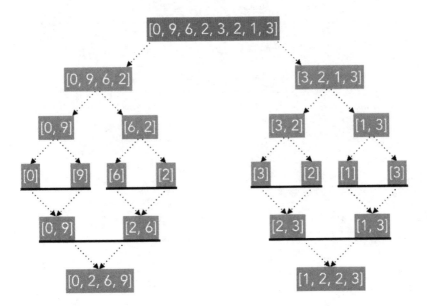

There is only one step left: during this last step, the two sorted halves are merged and sorted.

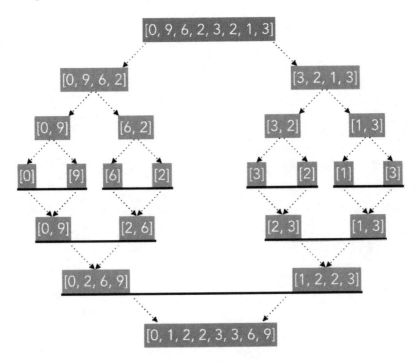

Finally, the result is the ordered array.

INTRODUCTION TO ALGORITHMS AND DATA STRUCTURES IN SWIFT 4

Now that you know how it works, let's implement this amazing algorithm.

Implementation

*If you want to follow along with me, download the repository from GitHub and open the MergeSort playground from the **advanced-sorting-src** folder.*

Here's the mergeSort() function:

```
func mergeSort(_ input: [Int]) -> [Int] {
    guard input.count > 1 else {
        return input
    }

    let splitIndex = input.count / 2
    let leftArray = mergeSort(Array(input[0..<splitIndex]))
    let rightArray = mergeSort(Array(input[splitIndex..<input.count]))

    return merge(leftPart: leftArray, rightPart: rightArray)
}
```

The input argument is the array of integers to be sorted. The function returns an array of ordered integers.

It doesn't make sense to sort an array which has one element.

```
guard input.count > 1 else {
    return input
}
```

The guard statement checks the size of the input array. If it has less than two elements, the function simply returns the input.

Next, we find the split index. We need to split the array in the middle. Therefore, the index is calculated as the input count over two.

As you probably noticed, we call the mergeSort() function from itself. This recursion is used to split the halves until one - or both - halves have only one element.

Then, the algorithm starts the sort-and-merge phase.

```
func merge(leftPart: [Int], rightPart: [Int]) -> [Int] {
    var sorted = [Int]()

    var leftIndex = 0
    var rightIndex = 0

    while leftIndex < leftPart.count && rightIndex < rightPart.count {
        if leftPart[leftIndex] < rightPart[rightIndex] {
            sorted.append(leftPart[leftIndex])
            leftIndex += 1
        } else if leftPart[leftIndex] > rightPart[rightIndex] {
            sorted.append(rightPart[rightIndex])
            rightIndex += 1
        } else {
            sorted.append(leftPart[leftIndex])
            leftIndex += 1
            sorted.append(rightPart[rightIndex])
            rightIndex += 1
        }
    }

    if leftIndex < leftPart.count {
        sorted.append(contentsOf: leftPart[leftIndex..<leftPart.count])
    } else if rightIndex < rightPart.count {
        sorted.append(contentsOf: rightPart[rightIndex..<rightPart.count])
    }

    return sorted
}
```

The merge(leftPart:, rightPart:) helper function compares the items from the left and right sublist.

```
            if leftPart[leftIndex] < rightPart[rightIndex] {
                sorted.append(leftPart[leftIndex])
                leftIndex += 1
            } else if leftPart[leftIndex] > rightPart[rightIndex] {
                sorted.append(rightPart[rightIndex])
```

```
        rightIndex += 1
    }
```

The smaller value is appended to the sorted array. If the original array has an odd number of elements, we cannot split it into exactly two halves. That's why we need the last checks.

```
    } else {
        sorted.append(leftPart[leftIndex])
        leftIndex += 1
        sorted.append(rightPart[rightIndex])
        rightIndex += 1
    }
```

The final step makes sure that the last element from either the left or the right subarray is added to the sorted list.

```
    if leftIndex < leftPart.count {
        sorted.append(contentsOf:
leftPart[leftIndex..<leftPart.count])
    } else if rightIndex < rightPart.count {
        sorted.append(contentsOf:
rightPart[rightIndex..<rightPart.count])
    }
```

> *The merge sort algorithm is a bit more challenging than the basic ones we've covered so far. I'd suggest you take a closer look at the playground project. Run it a couple of times, and try to understand its logic. Unfortunately, the playground does not come with a debugger. Still, you can print out the steps and the values to the console. Feel free to insert your print statements in the code wherever you need more clarity.*

So far we've talked about three basic and one more advanced sorting algorithm: selection sort, insertion sort, bubble sort and merge sort.

The merge sort is the most efficient among them. It has a logarithmic worst and average time complexity.

The merge sort uses a divide-and-conquer technique. It splits the original array recursively into smaller sublists. Then, in the sort and merge phase it constructs the ordered result.

Merge sort works best with larger inputs; if you need to sort tiny arrays - say, less than 20 elements - insertion sort may perform better due to its simplicity.

As an interesting fact: merge sort beats quicksort for list types where data can only be accessed sequentially, like in the case of linked lists. Yet another proof that understanding how algorithms work and when to apply them is imperative for writing efficient code.

SECTION 8.2
QUICKSORT

Quicksort is probably the most widely used sorting algorithm.

The quicksort is likely to run faster than any other, compare-based sorting algorithms in most cases. The algorithm was invented in 1960 and it's been consistently studied and refined over time. Hoare, the inventor of the algorithm, Dijkstra, Lomuto and others have been working on improving the efficiency of the quick sort even further.

The popularity of the quicksort algorithm is related to its performance. Besides, it's not too difficult to implement, and it works well with many different input types.

Quicksort uses a *divide-and-conquer* technique like the merge sort. However, the approach is different. Unlike for the merge sort, the final sorting of elements happens before the merge phase.

Let's visualize how this algorithm works.

Here's our unsorted array:

[0, 9, 6, 2, 3, 2, 1, 3]

First, we need to pick a pivot.

[0, 9, 6, 2, **3**, 2, 1, 3]

The algorithm uses this pivot to split the array into three parts: one list with all the elements that are smaller than the pivot, one with the elements that are equal to the pivot and one sublist that contains all the elements that are bigger than the pivot.

Splitting the array around the pivot is called partitioning.

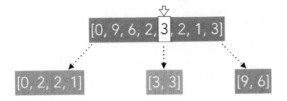

Next, we divide the leftmost sublist. As pivot, we choose the item at index 1.

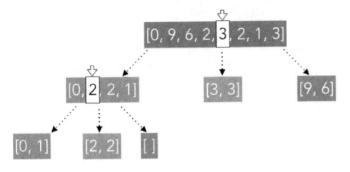

The resulting sublists contain the numbers [0, 1] and [2, 2].

Since there are no bigger numbers than 2, the sublist that should hold the bigger numbers is empty. Again, we pick a pivot, and split the next subarray: we get two arrays, each with only one element.

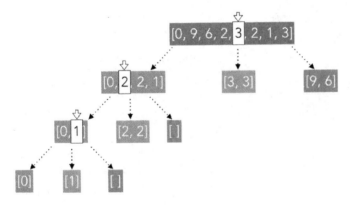

Now, we process the rightmost sublist which contains the elements that are greater than the very first pivot. We pick 6 as pivot, which gives us the following subarrays:

[] - Empty subarray for elements smaller than the pivot

[6] - Single element subarray for elements equal to the pivot

[9] - Single element subarray for elements bigger than the pivot

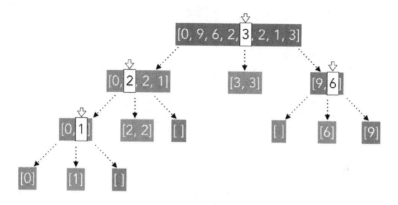

We are done with the partitioning phase.

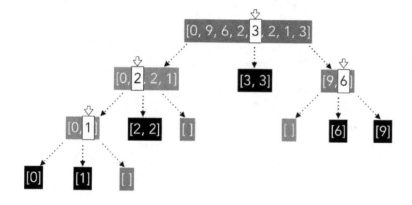

By merging the terminating sublists, we get the sorted result.

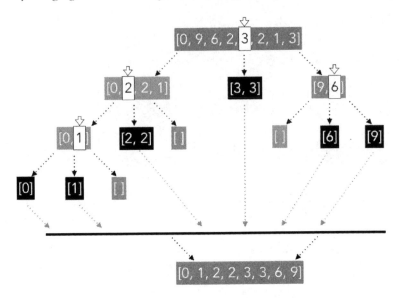

Now, let's implement the quicksort function.

Implementation

This quicksort variant is the simplest possible implementation.

> *If you want to follow along with me, download the repository from GitHub and open the Qsort playground from the **advanced-sorting-src** folder.*

```
func qsort(_ input: [Int]) -> [Int] {
    guard input.count > 1 else {
        return input
    }
    
    let pivotIndex = input.count / 2
    let pivot = input[pivotIndex]
    
    let less = input.filter {$0 < pivot}
    let equal = input.filter {$0 == pivot}
    let greater = input.filter {$0 > pivot}
    
    return qsort(less) + equal + qsort(greater)
}
```

The pivot is the element from the middle of the array.

```
let pivotIndex = input.count / 2
```

For partitioning, we rely on the filter() function.

The qsort() function calls itself recursively until all sublists contain one element or equal items. Finally, the sublists are combined. In Swift, you can simply use the add operator to concatenate arrays.

While this code is very simple, it's definitely not a production ready quicksort implementation. Instead of implementing a partitioning scheme, we rely on the filter() library function. This produces a very simple and compact code. However, this is not the best choice in terms of performance.

Many improvements have been made to the original quicksort algorithm, which was invented almost 60 years ago! Finding the optimal pivot and better partitioning schemes can further improve the efficiency of this clever algorithm.

A simple way to improve the performance of the quick sort is to switch to insertion sort for smaller arrays. Even Apple uses this trick in the implementation of the Swift sort function: https://github.com/apple/swift/blob/master/stdlib/public/core/Sort.swift.gyb

CHAPTER 9
WHERE DO YOU GO FROM HERE?

Congrats, you've reached the end of this book!

You've learned a lot about algorithms and you understand their benefits. Whenever in doubt, feel free to revisit the lectures in the section called "The Power of Algorithms."

The chapter about Big-O notation has clarified some of the most common time complexities through Swift code examples. Concepts like linear or quadratic time complexity won't make you raise your eyebrows anymore.

We delved into the details of three popular basic sorting algorithms and two advanced ones, including the extremely widespread quicksort. By now, you are probably able to explain and implement a sorting algorithm from scratch.

You should keep working on improving your algorithmic problem-solving skills. You'll have to practice a lot to make algorithmic thinking a habit. Instead of jumping to implementing a naive, slow solution, you'll eventually find yourself analyzing the problem and considering various aspects like worst-case or average time complexity and memory considerations.

You'll not only solve the problems, but you'll be able to provide elegant, efficient and reliable, long-term solutions.

SECTION 9.1
RESOURCES TO SHARPEN YOUR SKILLS

Now, you may want to deepen your knowledge further. So, what's next?

I give you some useful online resources which will help you in sharpening your coding and problem solving skills:

- Codility - https://www.codility.com

    It's a great resource for both developers and recruiters. It has many coding exercises and challenges to test your knowledge. The site provides an online editor and compiler, and supports a number of different programming languages, including Swift.
    You can provide custom test data and run several test rounds before submitting your solution. The solution is evaluated for correctness, edge-case scenarios and time complexity as well. You may not achieve the highest score even if your solution provides the expected results if it's performance is slow or it is fails some extreme edge case. An algorithmic approach is definitely required to solve most of the exercises on this site.

- Hackerrank - https://www.hackerrank.com - has a lot of tutorials and challenges.

- Project Euler - https://projecteuler.net - is a collection of challenging math and computer programming problems.

SECTION 9.2
GOODBYE!

I'd love to hear from you! Feel free to email me at carlos@leakka.com. And if you found this book useful, please leave a nice review or rating at https://www.amazon.com/dp/B077D8MQ31 .

Thank you!

SECTION 9.3
COPYRIGHT

© Copyright © 2018 by Károly Nyisztor.

All rights reserved. This book or any portion thereof may not be reproduced or used in any manner whatsoever without the express written permission of the publisher except for the use of brief quotations in a book review.

First Edition, 2018

Version 1.0

www.leakka.com

ABOUT THE AUTHOR

Károly Nyisztor is a software engineer, online instructor, and book author. His books are available on Amazon and iTunes. You can find his programming courses on Lynda, LinkedIn Learning, Udemy, Pluralsight and Skillshare.

He worked with companies such as Apple, Siemens, SAP, Zen Studios, and many more.
Károly has designed and built enterprise frameworks and several iOS apps and games, most of which were featured by Apple.

As an instructor, he shares his 20+ years of software development expertise and changes the lives of students throughout the world.
Károly is teaching software development related topics, including object-oriented software design, iOS programming, Swift, Objective-C, and UML.

Website: http://www.leakka.com
Youtube: https://www.youtube.com/c/swiftprogrammingtutorials
Twitter: https://twitter.com/knyisztor
Github: https://github.com/nyisztor

Udemy Courses: https://www.udemy.com/user/karolynyisztor
Pluralsight Courses: https://www.pluralsight.com/profile/author/karoly-nyisztor
Lynda Courses: https://www.lynda.com/Karoly-Nyisztor/9655357-1.html

Made in the USA
San Bernardino, CA
18 March 2018